LIGHT

TASK CARD SERIES

Conceived and written by
RON MARSON

Illustrated by
PEG MARSON

TOPS LEARNING SYSTEMS

342 S Plumas Street
Willows, CA 95988

www.topscience.org

WHAT CAN YOU COPY?

Dear Educator,

Please honor our copyright restrictions. We offer liberal options and guidelines below with the intention of balancing your needs with ours. When you buy these labs and use them for your own teaching, you sustain our work. If you "loan" or circulate copies to others without compensating TOPS, you squeeze us financially, and make it harder for our small non-profit to survive. Our well-being rests in your hands. Please help us keep our low-cost, creative lessons available to students everywhere. Thank you!

PURCHASE, ROYALTY and LICENSE OPTIONS

TEACHERS, HOMESCHOOLERS, LIBRARIES:

We do all we can to keep our prices low. Like any business, we have ongoing expenses to meet. We trust our users to observe the terms of our copyright restrictions. While we prefer that all users purchase their own TOPS labs, we accept that real-life situations sometimes call for flexibility.

Reselling, trading, or loaning our materials is prohibited unless one or both parties contribute an Honor System Royalty as fair compensation for value received. We suggest the following amounts – let your conscience be your guide.

HONOR SYSTEM ROYALTIES: If making copies from a library, or sharing copies with colleagues, please calculate their value at 50 cents per lesson, or 25 cents for homeschoolers. This contribution may be made at our website or by mail (addresses at the bottom of this page). Any additional tax-deductible contributions to make our ongoing work possible will be accepted gratefully and used well.

Please follow through promptly on your good intentions. Stay legal, and do the right thing.

SCHOOLS, DISTRICTS, and HOMESCHOOL CO-OPS:

PURCHASE Option: Order a book in quantities equal to the number of target classrooms or homes, and receive quantity discounts. If you order 5 books or downloads, for example, then you have unrestricted use of this curriculum for any 5 classrooms or families per year for the life of your institution or co-op.

2-9 copies of any title: 90% of current catalog price + shipping.

10+ copies of any title: 80% of current catalog price + shipping.

ROYALTY/LICENSE Option: Purchase just one book or download *plus* photocopy or printing rights for a designated number of classrooms or families. If you pay for 5 additional Licenses, for example, then you have purchased reproduction rights for an entire book or download edition for any **6** classrooms or families per year for the life of your institution or co-op.

1-9 Licenses: 70% of current catalog price per designated classroom or home.

10+ Licenses: 60% of current catalog price per designated classroom or home.

WORKSHOPS and TEACHER TRAINING PROGRAMS:

We are grateful to all of you who spread the word about TOPS. Please limit copies to only those lessons you will be using, and collect all copyrighted materials afterward. No take-home copies, please. Copies of copies are strictly prohibited.

ISBN 978 - 0 - 941008-87-7

CONTENTS

A TOPS Model for Effective Science Teaching...

If science were only a set of explanations and a collection of facts, you could teach it with blackboard and chalk. You could assign students to read chapters and answer the questions that followed. Good students would take notes, read the text, turn in assignments, then give you all this information back again on a final exam. Science is traditionally taught in this manner. Everybody learns the same body of information at the same time. Class togetherness is preserved.

But science is more than this.

Science is also process — a dynamic interaction of rational inquiry and creative play. Scientists probe, poke, handle, observe, question, think up theories, test ideas, jump to conclusions, make mistakes, revise, synthesize, communicate, disagree and discover. Students can understand science as process only if they are free to think and act like scientists, in a classroom that recognizes and honors individual differences.

Science is *both* a traditional body of knowledge *and* an individualized process of creative inquiry. Science as process cannot ignore tradition. We stand on the shoulders of those who have gone before. If each generation reinvents the wheel, there is no time to discover the stars. Nor can traditional science continue to evolve and redefine itself without process. Science without this cutting edge of discovery is a static, dead thing.

Here is a teaching model that combines the best of both elements into one integrated whole. It is only a model. Like any scientific theory, it must give way over time to new and better ideas. We challenge you to incorporate this TOPS model into your own teaching practice. Change it and make it better so it works for you.

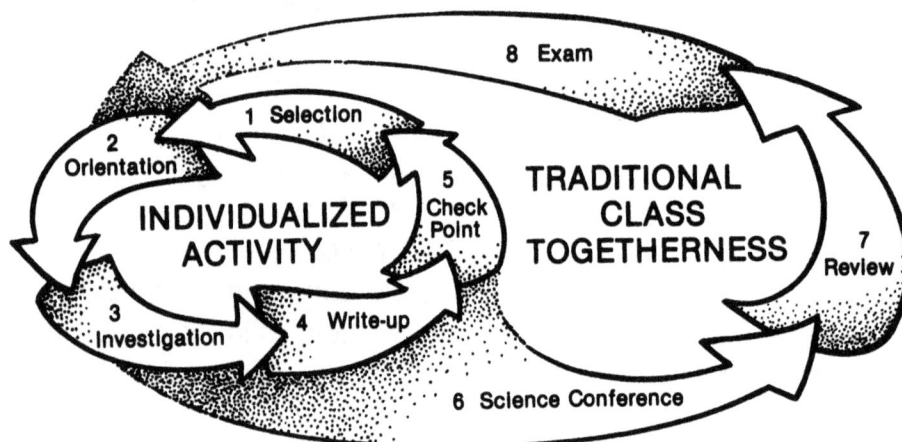

1. SELECTION

Doing TOPS is as easy as selecting the first task card and doing what it says, then the second, then the third, and so on. Working at their own pace, students fall into a natural routine that creates stability and order. They still have questions and problems, to be sure, but students know where they are and where they need to go.

Students generally select task cards in sequence because new concepts build on old ones in a specific order. There are, however, exceptions to this rule: students might *skip* a task that is not challenging; *repeat* a task with doubtful results; *add* a task of their own design to answer original "what would happen if" questions.

2. ORIENTATION

Many students will simply read a task card and immediately understand what to do. Others will require further verbal interpretation. Identify poor readers in your class. When they ask, "What does this mean?" they may be asking in reality, "Will you please read this card aloud?"

With such a diverse range of talent among students, how can you individualize activity and still hope to finish this module as a cohesive group? It's easy. By the time your most advanced students have completed all the task cards, including the enrichment series at the end, your slower students have at least completed the basic core curriculum. This core provides the common background so necessary for meaningful discussion, review and testing on a class basis.

3. INVESTIGATION

Students work through the task cards independently and cooperatively. They follow their own experimental strategies and help each other. You encourage this behavior by helping students only *after* they have tried to help themselves. As a resource person, you work to stay *out* of the center of attention, answering student questions rather than posing teacher questions.

When you need to speak to everyone at once, it is appropriate to interrupt individual task card activity and address the whole class, rather than repeat yourself over and over again. If you plan ahead, you'll find that most interruptions can fit into brief introductory remarks at the beginning of each new period.

4. WRITE-UP

Task cards ask students to explain the "how and why" of things. Write-ups are brief and to the point. Students may accelerate their pace through the task cards by writing these reports out of class.

Students may work alone or in cooperative lab groups. But each one must prepare an original write-up. These must be brought to the teacher for approval as soon as they are completed. Avoid dealing with too many write-ups near the end of the module, by enforcing this simple rule: each write-up must be approved *before* continuing on to the next task card.

5. CHECK POINT

The student and teacher evaluate each write-up together on a pass/no-pass basis. (Thus no time is wasted haggling over grades.) If the student has made reasonable effort consistent with individual ability, the write-up is checked off on a progress chart and included in the student's personal assignment folder or notebook kept on file in class.

Because the student is present when you evaluate, feedback is immediate and effective. A few seconds of this direct student-teacher interaction is surely more effective than 5 minutes worth of margin notes that students may or may not heed. Remember, you don't have to point out every error. Zero in on particulars. If reasonable effort has not been made, direct students to make specific improvements, and see you again for a follow-up check point.

A responsible lab assistant can double the amount of individual attention each student receives. If he or she is mature and respected by your students, have the assistant check the even-numbered write-ups while you check the odd ones. This will balance the work load and insure that all students receive equal treatment.

6. SCIENCE CONFERENCE

After individualized task card activity has ended, this is a time for students to come together, to discuss experimental results, to debate and draw conclusions. Slower students learn about the enrichment activities of faster students. Those who did original investigations, or made unusual discoveries, share this information with their peers, just like scientists at a real conference. This conference is open to films, newspaper articles and community speakers. It is a perfect time to consider the technological and social implications of the topic you are studying.

7. READ AND REVIEW

Does your school have an adopted science textbook? Do parts of your science syllabus still need to be covered? Now is the time to integrate other traditional science resources into your overall program. Your students already share a common background of hands-on lab work. With this shared base of experience, they can now read the text with greater understanding, think and problem-solve more successfully, communicate more effectively.

You might spend just a day on this step or an entire week. Finish with a review of key concepts in preparation for the final exam. Test questions in this module provide an excellent basis for discussion and study.

8. EXAM

Use any combination of the review/test questions, plus questions of your own, to determine how well students have mastered the concepts they've been learning. Those who finish your exam early might begin work on the first activity in the next new TOPS module.

Now that your class has completed a major TOPS learning cycle, it's time to start fresh with a brand new topic. Those who messed up and got behind don't need to stay there. Everyone begins the new topic on an equal footing. This frequent change of pace encourages your students to work hard, to enjoy what they learn, and thereby grow in scientific literacy.

GETTING READY

Here is a checklist of things to think about and preparations to make before your first lesson.

☐ Decide if this TOPS module is the best one to teach next.

TOPS modules are flexible. They can generally be scheduled in any order to meet your own class needs. Some lessons within certain modules, however, do require basic math skills or a knowledge of fundamental laboratory techniques. Review the task cards in this module now if you are not yet familiar with them. Decide whether you should teach any of these other TOPS modules first: *Measuring Length, Graphing, Metric Measure, Weighing* or *Electricity* (before *Magnetism*). It may be that your students already possess these requisite skills or that you can compensate with extra class discussion or special assistance.

☐ Number your task card masters in pencil.

The small number printed in the lower right corner of each task card shows its position within the overall series. If this ordering fits your schedule, copy each number into the blank parentheses directly above it at the top of the card. Be sure to use pencil rather than ink. You may decide to revise, upgrade or rearrange these task cards next time you teach this module. To do this, write your own better ideas on blank 4 x 6 index cards, and renumber them into the task card sequence wherever they fit best. In this manner, your curriculum will adapt and grow as you do.

☐ Copy your task card masters.

You have our permission to reproduce these task cards, for as long as you teach, with only 1 restriction: please limit the distribution of copies you make to the students you personally teach. Encourage other teachers who want to use this module to purchase their *own* copy. This supports TOPS financially, enabling us to continue publishing new TOPS modules for you. For a full list of task card options, please turn to the first task card masters numbered "cards 1-2."

☐ Collect needed materials.

Please see the opposite page.

☐ Organize a way to track completed assignment.

Keep write-ups on file in class. If you lack a vertical file, a box with a brick will serve. File folders or notebooks both make suitable assignment organizers. Students will feel a sense of accomplishment as they see their file folders grow heavy, or their notebooks fill up, with completed assignments. Easy reference and convenient review are assured, since all papers remain in one place.

Ask students to staple a sheet of numbered graph paper to the inside front cover of their file folder or notebook. Use this paper to track each student's progress through the module. Simply initial the corresponding task card number as students turn in each assignment.

☐ Review safety procedures.

Most TOPS experiments are safe even for small children. Certain lessons, however, require heat from a candle flame or Bunsen burner. Others require students to handle sharp objects like scissors, straight pins and razor blades. These task cards should not be attempted by immature students unless they are closely supervised. You might choose instead to turn these experiments into teacher demonstrations.

Unusual hazards are noted in the teaching notes and task cards where appropriate. But the curriculum cannot anticipate irresponsible behavior or negligence. It is ultimately the teacher's responsibility to see that common sense safety rules are followed at all times. Begin with these basic safety rules:

1. Eye Protection: Wear safety goggles when heating liquids or solids to high temperatures.
2. Poisons: Never taste anything unless told to do so.
3. Fire: Keep loose hair or clothing away from flames. Point test tubes which are heating away from your face and your neighbor's.
4. Glass Tubing: Don't force through stoppers. (The teacher should fit glass tubes to stoppers in advance, using a lubricant.)
5. Gas: Light the match first, before turning on the gas.

☐ Communicate your grading expectations.

Whatever your philosophy of grading, your students need to understand the standards you expect and how they will be assessed. Here is a grading scheme that counts individual effort, attitude and overall achievement. We think these 3 components deserve equal weight:

1. Pace (effort): Tally the number of check points you have initialed on the graph paper attached to each student's file folder or science notebook. Low ability students should be able to keep pace with gifted students, since write-ups are evaluated relative to individual performance standards. Students with absences or those who tend to work at a slow pace may (or may not) choose to overcome this disadvantage by assigning themselves more homework out of class.

2. Participation (attitude): This is a subjective grade assigned to reflect each student's attitude and class behavior. Active participators who work to capacity receive high marks. Inactive onlookers, who waste time in class and copy the results of others, receive low marks.

3. Exam (achievement): Task cards point toward generalizations that provide a base for hypothesizing and predicting. A final test over the entire module determines whether students understand relevant theory and can apply it in a predictive way.

Gathering Materials

Listed below is everything you'll need to teach this module. You already have many of these items. The rest are available from your supermarket, drugstore and hardware store. Laboratory supplies may be ordered through a science supply catalog.

Keep this classification key in mind as you review what's needed:

special in-a-box materials:	general on-the-shelf materials:
Italic type suggests that these materials are unusual. Keep these specialty items in a separate box. After you finish teaching this module, label the box for storage and put it away, ready to use again the next time you teach this module.	Normal type suggests that these materials are common. Keep these basics on shelves or in drawers that are readily accessible to your students. The next TOPS module you teach will likely utilize many of these same materials.
(substituted materials):	*optional materials:
Parentheses enclosing any item suggests a ready substitute. These alternatives may work just as well as the original, perhaps better. Don't be afraid to improvise, to make do with what you have.	An asterisk sets these items apart. They are nice to have, but you can easily live without them. They are probably not worth an extra trip to the store, unless you are gathering other materials as well.

Everything is listed in order of first use. Start gathering at the top of this list and work down. Ask students to bring recycled items from home. The teaching notes may occasionally suggest additional student activity under the heading "Extensions." Materials for these optional experiments are listed neither here nor in the teaching notes. Read the extension itself to find out what new materials, if any, are required.

Needed quantities depend on how many students you have, how you organize them into activity groups, and how you teach. Decide which of these 3 estimates best applies to you, then adjust quantities up or down as necessary:

$Q_1 / Q_2 / Q_3$

Single Student: Enough for 1 student to do all the experiments.
Individualized Approach: Enough for 30 students informally working in 10 lab groups, all self-paced.
Traditional Approach: Enough for 30 students, organized into 10 lab groups, all doing the same lesson.

KEY: *special in-a-box materials* *(substituted materials)*	general on-the-shelf materials *optional materials

1/1/1	*packet powdered milk (fresh milk)*	5/50/50	straight plastic drinking straws
1/1/1	source of water	2/6/20	*bathroom hand mirrors with plane surfaces*
1/10/10	baby food jars	4/40/40	microscope slides
1/10/10	*dark-colored cloth towels*	1/6/10	*rectangular cake pans — about 9 by 12 inches or 20 x 30 cm*
1/10/10	flashlights with two fresh D cells — ask students to bring these from home.	1/6/10	eyedroppers
1/1/1	roll aluminum foil.	1/4/10	empty tuna fish cans or equivalent size
1/6/10	rolls masking tape	2/8/20	small test tubes
2/20/20	*rectangular pocket mirrors without frames*	4/40/40	hand lenses with rigid handles — see notes 18 for recommendations
2/20/20	empty tin cans — medium 15 or 16 oz size	1/4/10	*transparent plastic pill vials — see notes 19
1/1/1	roll waxed paper	1/4/10	*prisms — see notes 20*
4/40/40	medium-sized rubber bands	1/4/10	white plates (paper plates)
3/30/30	straight pins	1/1/1	*sheets each of blue and yellow cellophane (clear plastic report covers)*
1/10/10	scissors		
1/10/10	*empty cereal boxes*	2/20/20	textbooks of equal size
3/30/30	size-D batteries, dead or alive	1/1/1	roll plastic wrap
5/50/50	4x6 inch index cards	1/4/10	ball point pens
5/50/50	paper clips	1/5/10	rolls clear tape
1/2/5	paper punch tools	1/6/10	canning rings
1/1/1	spool thread	1/4/10	spoons
2/20/20	pennies	1/10/10	facial tissues (soft toilet tissues)
10/100/100	sheets notebook paper		

Sequencing Task Cards

This logic tree shows how all the task cards in this module tie together. In general, students begin at the trunk of the tree and work up through the related branches. As the diagram suggests, the way to upper level activities leads up from lower level activities.

At the teacher's discretion, certain activities can be omitted or sequences changed to meet specific class needs. The only activities that must be completed in sequence are indicated by leaves that open *vertically* into the ones above them. In these cases the lower activity is a prerequisite to the upper.

When possible, students should complete the task cards in the same sequence as numbered. If time is short, however, or certain students need to catch up, you can use the logic tree to identify concept-related *horizontal* activities. Some of these might be omitted since they serve only to reinforce learned concepts rather than introduce new ones.

On the other hand, if students complete all the activities at a certain horizontal concept level, then experience difficulty at the next higher level, you might go back down the logic tree to have students repeat specific key activities for greater reinforcement.

For whatever reason, when you wish to make sequence changes, you'll find this logic tree a valuable reference. Parentheses in the upper right corner of each task card allow you total flexibility. They are left blank so you can pencil in sequence numbers of your own choosing.

LIGHT 17

E

LONG-RANGE OBJECTIVES

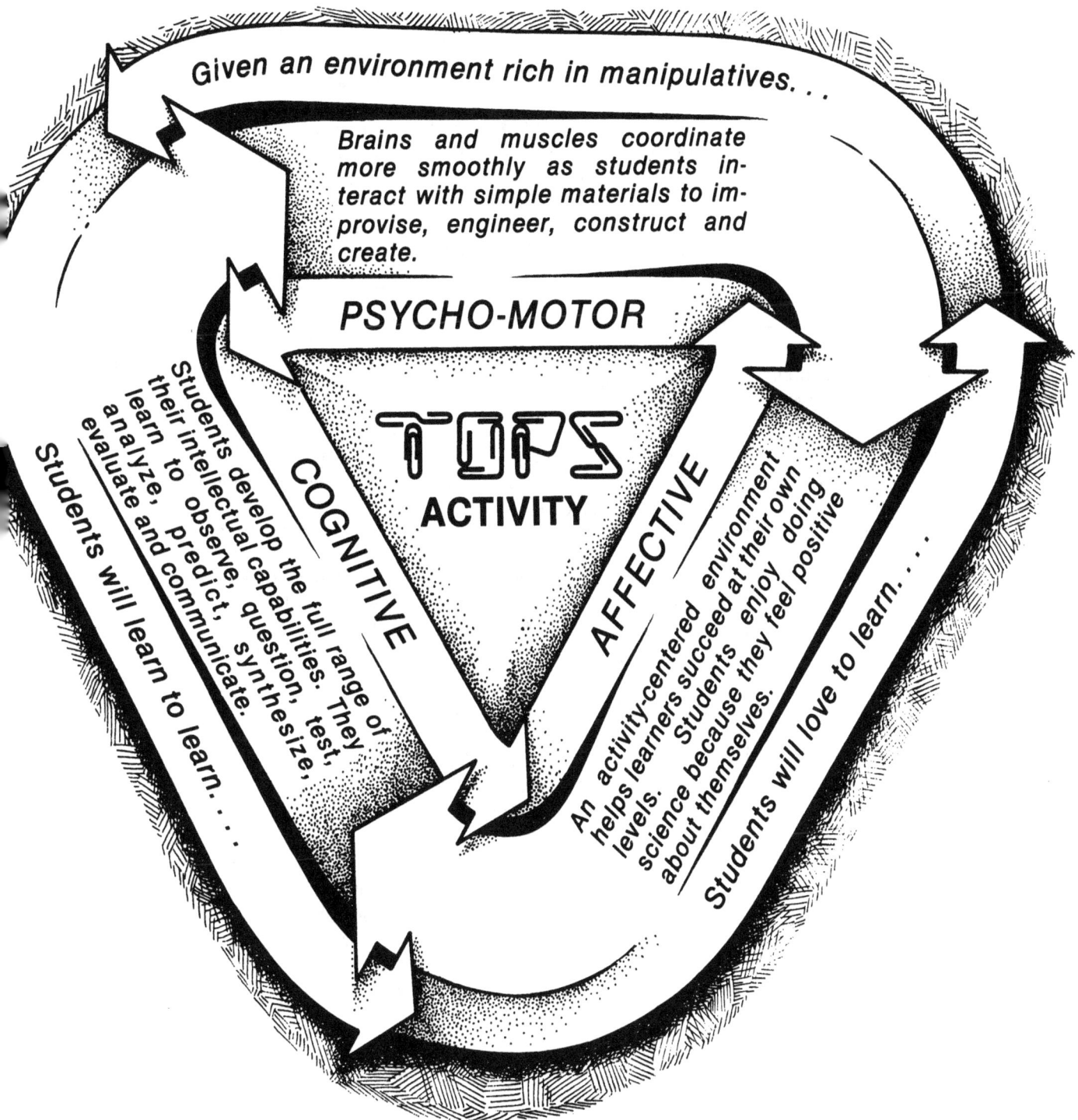

Given an environment rich in manipulatives. . .

Brains and muscles coordinate more smoothly as students interact with simple materials to improvise, engineer, construct and create.

PSYCHO-MOTOR

TOPS ACTIVITY

COGNITIVE

Students develop the full range of their intellectual capabilities. They learn to observe, question, test, analyze, predict, synthesize, evaluate and communicate.

AFFECTIVE

An activity-centered environment helps learners succeed at their own levels. Students enjoy doing science because they feel positive about themselves.

Students will learn to learn. . . . .

Students will love to learn. . . . .

Review / Test Questions

Photocopy the questions below. On a separate sheet of blank paper, cut and paste those boxes you want to use as test questions. Include questions of your own design, as well. Crowd all these questions onto a single page for students to answer on another paper, or leave space for student responses after each question, as you wish. Duplicate a class set and your custom-made test is ready to use. Use leftover questions as a review in preparation for the final exam.

task 1
How does a pea shooter (a straw through which you blow peas) model a flashlight?

task 2-3 A
Looking directly at the sun (during an eclipse or any other time) can cause irreversible damage to the retina of your eye. Explain how to use a pinhole to safely watch the sun.

task 2-3 B
As you change the distance between a pinhole viewer and an object, how does the size of the image change? Illustrate your answer with diagrams.

task 4-5 A
The shadow of the monster looms larger and larger against the castle wall. How is this monster moving relative to the wall and the light source? Use diagrams to support your answer.

task 4-5 B
Which projects a clearer shadow of your hand against a wall — a small pen light or a large lantern? Use the terms umbra and penumbra in your answer.

task 6
You are observing (in a safe manner) an eclipse of the sun. Draw how the sun appears if you…
a. stand in twilight.
b. stand in total darkness.
c. notice almost no darkness at all.

task 7
You are adrift in a tiny raft at sea. Just before sunset you spot a search plane high overhead. You grab a hand mirror from your emergency pack and feverishly try to signal up an SOS. How should you hold the mirror?

task 7-10
Accurately sketch the virtual image of this arrow behind the mirror. Explain how your drawing illustrates that $\angle i = \angle r$.

task 11
Can the observer see each point x, y and z reflected in the mirror? Illustrate your answer with a diagram.

task 12
How could you use a piece of glass to create the illusion of a candle flame burning inside a jar of water?

task 13
How does water dripping into a full tub model a light bulb shining in a room?

task 14-16
Ocean waves may move in a variety of directions far from land, but generally form wave fronts parallel to the shore when near land.

a. Explain why this happens.
b. How does this model the refraction of light?

task 7, 16
Trace the path of this ray as it refracts between air and water and reflects off the mirror.

task 17
Complete this diagram to explain why each point at the bottom of a swimming pool appears to be more shallow than it really is.

task 18, 24
Show how sunlight passes through each lens. Label the concave lens, the convex lens, converging rays, diverging rays and the focal point.

Answers

task 1
Switching on the flashlight is analogous to shooting peas. Tiny photons of light, like peas, move out from the bulb, in a straight line, until they strike a wall, mirror, etc., and bounce (reflect) in a new direction.

task 2-3 A
Remove both ends from a can. Cover one end with foil and poke a pinhole in it; screen the other end with waxed paper. Now point the pinhole at the sun. Its light will project through the hole onto the screen, where it can be safely viewed. For best results, darken the screen with your hands or shield it with a second can.

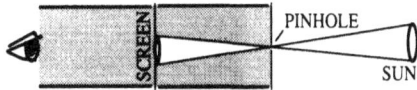

Two index cards, one with a pinhole held in front of another serving as a screen, will also work. The shadow cast by the first card provides good contrast for the projected pinhole image on the second.

task 2-3 B
The closer an object is to the pinhole, the larger it appears on the screen.

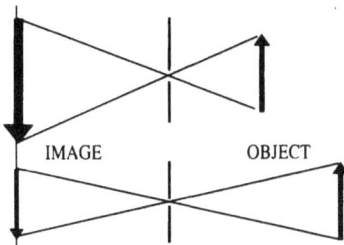

task 4-5 A
The monster is moving away from the wall, toward a source of light that is relatively near. If the light was far away, with the rays nearly parallel, the shadow would not grow.

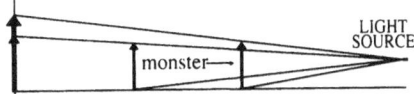

task 4-5 B
The pen light forms the clearer image. Because all its light originates from a single point, it casts a pure dark umbra behind the hand.

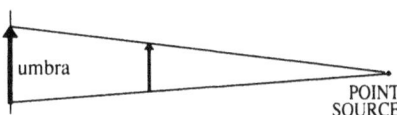

The lantern, by contrast, casts light from many points scattered over a much wider area. This creates a complex penumbra around the outside perimeter of the shadow that is illuminated by some points on the lantern, but not others.

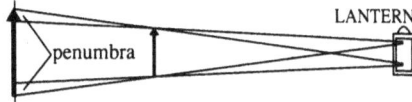

task 6

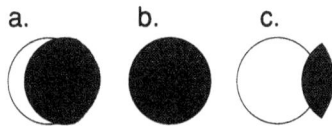

task 7
Hold the mirror at about 45°, so incident light approaching horizontally from the setting sun reflects vertically, through an equal angle from the normal.

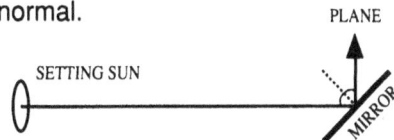

task 7-10
Each incident light ray reflects off the mirror through an equal angle as measured from the normal. These reflected rays trace back to common points that define the image.

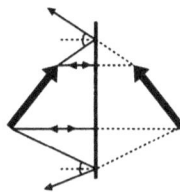

task 11
The observer sees points x and y, because light from these positions reflects through equal angles of incidence and reflection to reach the observers eye. The mirror is too high for rays from point z to reflect to the eye.

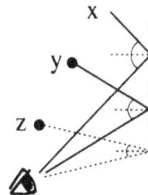

task 12
Position the glass vertically, halfway between a candle in front and a jar of water behind. When you light the candle, the virtual image of its flame reflected off the glass will super-impose over the real image of the jar of water seen through the glass.

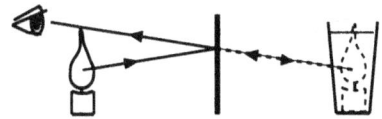

task 13
The water waves move outward in all directions from each drop, then reflect off the sides of the tub. In a similar manner, light moves outward from the bulb in all directions, then reflects off the walls of the room.

task 14-16
a. Surface waves tend to slow down as they interact with the rising ocean floor near the shoreline. This drag causes faster parts of the wave still in deep water to pivot around the slower parts, turning the wave directly toward the shore.

b. In a similar manner, light travels slower through mediums of greater density. When a light beam passes between air and glass, for example, that part of the wave front that first enters the glass moves slower than the remainder still in air. This pivots the wave toward the glass just as an ocean wave pivots towards shore.

task 7, 16

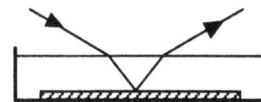

task 17

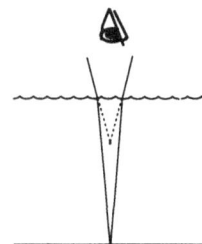

task 18, 24

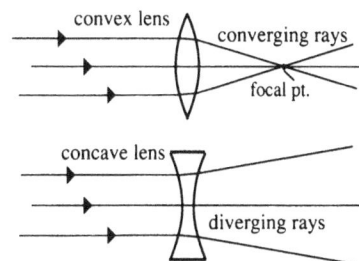

Review / Test Questions (continued)

task 18, 24, 30
Which of these lenses...
a. has no focal point?
b. has the longest focal length?
c. has the shortest focal length?
d. magnifies the most?

w.

x.

y.

z.

task 19
Diagram at least 2 points of refraction and 2 points of reflection for the incident light ray.

INCIDENT
RAY

GLASS PRISM

task 20
The same spot on a diamond ring flashes many different colors as you gently rock it back and forth on your finger in sunlight. How is this possible?

task 20-21
Is white a real color in the same sense that red and blue are real colors?

task 21
When *paint* is mixed:
    blue + yellow = green.
When *light* is mixed:
    blue + yellow = white.
How can this be?

task 22-27
What is the difference between...
a. a concave lens and a convex lens?
b. a focal point and a focal length?
c. a real image and a virtual image?

task 22-23
You are given a magnifying glass. Describe 2 different ways to measure its focal length.

task 24
For each incident ray below, describe its emerging path on the opposite side of the lens.

v.

w. focal point

x.

y.

focal point z.

task 24-26
Use ray diagrams to find the image of each object arrow.

f   f

f   f

task 25-27
Where must an object be located so the image formed by a hand lens is...
a. virtual
b. real
c. inverted and enlarged
d. inverted and reduced
e. inverted, but the same size

task 28-29 A
Does a microscope form a virtual image? A real image? Explain.

task 28-29 B
How is a telescope similar to a microscope? How are they different?

task 30
Look at something far away, then at something close. What does your eye do to allow you to see both images clearly? Illustrate your answer with diagrams.

task 31-32
How is a tiny pinhole poked in foil like a hand lens? How is it different?

task 33
You are assigned to paint the word FIRE on the front of a truck so other drivers can recognize the word when looking into their rearview mirrors. Draw how it should look. Use a mirror if you want to.

task 34
Draw in all lines of symmetry for each figure. Use a mirror if you wish.

task 35-36
A hand lens reflects back 2 images of a candle flame.
a. What part of the lens forms each image?
b. Which image is virtual and which is real? How could you prove your answer?

Answers (continued)

task 18, 24, 30
a. lens x has no focal point.
b. lens z has the longest focal length.
c. lens y has the shortest focal length.
d. lens y magnifies the most.

task 19

task 20
The diamond refracts sunlight into a spectrum of its component colors because different wavelengths refract through different angles. Blue bends the most, red the least. These colors, and all others between, enter the eye as separate light rays.

task 20-21
No. White is a mixture of all colors encompassing the whole visible spectrum, including red and blue.

task 21
Mixing paint *subtracts* color: blue absorbs the shorter wavelengths while yellow absorbs the longer wavelengths, leaving only green (in the middle range) to be reflected back from the incident white light. Mixing light, by contrast, *adds* color: the shorter blue-end wavelengths add to longer yellow-end wave-lengths, presenting the entire spectrum, which you see as white.

task 22-27
a. A concave lens curves inward while a convex lens bulges outward.
b. Parallel incident rays converge at a specific *focal point* in space. The distance from this point to the center of the lens is known as the *focal length*.
c. A real image appears opposite its object, on the other side of the lens. It can be projected onto a screen. A virtual image appears on the same side of the lens as its object. It cannot be projected onto a screen.

task 22-23
Method 1: Sight through the lens to lines that run parallel to the principal axis. These will appear to converge at the focal point and can be drawn by sighting along a straight edge. Measure the distance from where these lines intersect to the center of the lens, along the principal axis.

Method 2: Focus light from any bright distant object on a white screen (an index card). The object will come into sharpest focus one focal length away from the lens.

task 24
The emerging ray...
v. passes through the focal point.
w. runs parallel to the principal axis.
x. continues along the same straight line as the incident ray.
y. diverges from the principal axis.
z. converges with the principal axis.

task 24-26

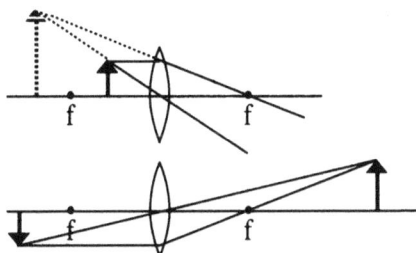

task 25-27
The object must be located...
a. inside 1 f.l.
b. outside 1 f.l.
c. outside 1 f.l. but inside 2 f.l.
d. outside 2 f.l.
e. at 2 f.l.

task 28-29 A
A microscope forms both kinds of images. As light enters the bottom lens (the objective) it forms a real inverted image. This image is then magnified by the top lens (the eye piece) to form a virtual image.

task 28-29 B
Similarities: Both instruments form images of images using 2 or more lenses. In 2-lens instruments, the first image, which is inverted and real, is then magnified to an enlarged virtual image by the second lens.
Differences: A telescope focuses on distant objects, forming a reduced, real, inverted image just beyond the focal length of the front lens. A microscope focuses on close objects, forming an enlarged, real, inverted image well beyond 2 focal lengths of the front lens.

task 30
The eye increases the curvature of its lens when looking at closer objects. This increases the diffraction of the more divergent incoming rays so they can still focus on the retina.

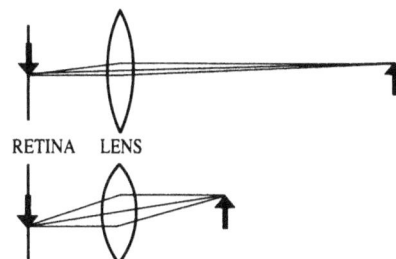

RETINA   LENS

task 31-32
Similar: Both focus light from an object to form a real, inverted image on a screen.
Different: The pinhole has no focal point like a lens, forming instead a real image at all distances from the screen. The pinhole doesn't refract light to a focal point like a lens, rather it blocks out all unfocused light, admitting only a single ray (ideally) from each point on the object to the screen. Nor can the pinhole form erect, virtual images like a lens.

task 33

task 34

task 35-36
a. The flame image in front of the lens reflects off the back, concave surface of the lens, while the image in back reflects off its front, convex surface.
b. The "concave" reflection projected to the front is real. The flame can be captured on a paper screen held in front of the lens. The "convex" reflection appearing behind the lens is virtual. Place a hand over the back of the lens, and the flame appears to burn from inside your hand. It is not really where it appears to be, nor can it be captured on a screen.

TEACHING NOTES
For Activities 1-36

Task Objective (TO) observe the path of light beams in a colloidal suspension. To understand light beams in terms of the particle nature of light.

LIGHT AS PARTICLES　　　　O　　　　　　Light ( )

1. Add a *tiny* pinch of powdered milk to a small jar of water. Shake it vigorously. It should look only slightly cloudy.

SLIGHTLY CLOUDY SUSPENSION

2. Wrap the jar in a dark cloth while shining a flashlight through a small opening. Look through the top and record your observations.

DARK CLOTH

3. Tape a piece of foil to the jar, shiny side in. Repeat the experiment, directing the beam through the water at the foil. What do you see?

TAPE

FOIL (Shiny side in)

4. It is useful to think of light as composed of incredibly tiny, fast moving particles (*photons*). How do these photons appear to move?

PHOTON

RAY

5. The path a light photon takes is called a *ray*. Draw a light ray diagram to illustrate why you can't see your ear without a mirror.

© 1991 by TOPS Learning Systems

1

Answers / Notes

1. *The water should look only slightly cloudy, remaining essentially transparent.*

2. A straight beam of light cuts through the milky suspension.

3. The light beam reflects off the foil, deflecting along a new straight path.

4. Photons seem to move along straight lines. When they hit shiny aluminum foil they are able to bounce off, reflecting along a new straight line. *(The particle nature of light presented here is balanced by a presentation of light as waves in activity 13.)*

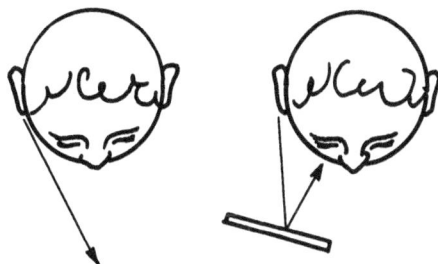

5.

Materials

☐ Powdered milk. A drop of fresh milk will also serve.
☐ A source of water.
☐ A small beaker or jar. Baby food jars, used widely in this module, are ideal.
☐ A dark-colored towel or cloth.
☐ A flashlight.
☐ Aluminum foil.
☐ Masking tape.
☐ A plane mirror (optional).

(TO) view an image through a pinhole. To draw a ray diagram that explains why pinhole images are inverted.

PINHOLE VIEWER (1)　　　　○　　　　　　　Light (　　)

1. Get 2 tin cans of equal size with both ends removed. Cover just one with foil at one end, and rubber band waxed paper to the other end.

RUBBER BAND

SECOND CAN UNCOVERED

FOIL

WAXED PAPER

2. Poke a pinhole in the center of the foil. Enlarge it to the size of a pin head with a sharp pencil point.

POKE

ENLARGE TO PINHEAD SIZE

3. Hold the uncovered can over the waxed paper end, while looking through it toward a well-lit area.

    a. Write your observations.
    b. Is the pinhole image erect (right-side-up) or inverted (upside-down)?
    c. Draw a labeled light ray diagram to explain your observations.

Save your pinhole viewer to use again.

COVERED CAN　　UNCOVERED CAN

PINHOLE

2

Introduction

    Diagrams serve a scientific purpose. They explain but do not give real representations. Light, for example, does not consist of straight lines, nor do arrows represent an object or image as an artist would draw them. Yet straight lines and arrows do help us interpret our world. When students explain things in pictures for the first time, they tend to draw like artists, including details that obscure rather than illustrate. Demonstrate the simple direct nature of diagrams using examples like these:

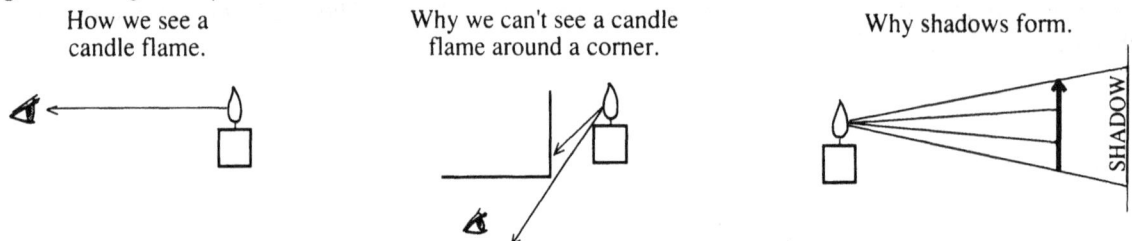

How we see a candle flame.

Why we can't see a candle flame around a corner.

Why shadows form.

SHADOW

Answers / Notes

3. *This pinhole viewer can also be used to safely view the sun. Remind students, however, that the unshielded eye can be irreversibly damaged by looking directly at the sun.*
3a. Recognizable images form on the waxed paper.
3b. The pinhole images are inverted.

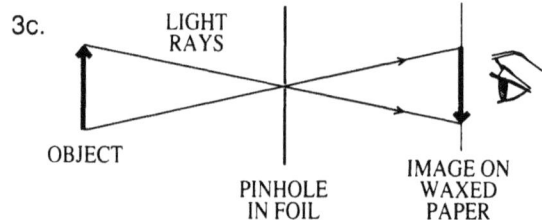

3c.

LIGHT RAYS

OBJECT

PINHOLE IN FOIL

IMAGE ON WAXED PAPER

Materials

☐ Two cans with both ends removed. Use medium sized cans. Fifteen ounce vegetable cans are an ideal size to use here and in other experiments.
☐ Aluminum foil and waxed paper.
☐ A rubber band.

☐ A straight pin.
☐ A sharpened pencil.
☐ A bright light source. If you don't have outside windows, or bright incandescent lights, supply a candle, matches and close supervision.

(TO) observe how a pinhole image changes position and size. To explain these variations using ray diagrams.

PINHOLE VIEWER (2)   O   Light (   )

1. Look at a shining flashlight through your pinhole viewer. Use ray diagrams to explain each observation.

   a. As you move the object close or far away, the pinhole image grows or shrinks.

   b. As you move the object in one direction, the pinhole image moves in the opposite direction.

IMAGE

OBJECT

2. Think of a way to put multiple images of the same object on your screen. Explain how you did this.

© 1991 by TOPS Learning Systems

3

Answers / Notes

1. *Good observers may notice, at certain angles, the looping pattern of the glowing bulb filament within the larger illuminated circle of the flashlight head.*

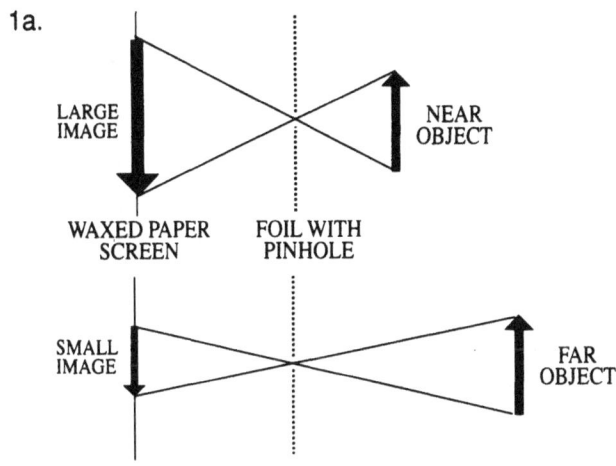

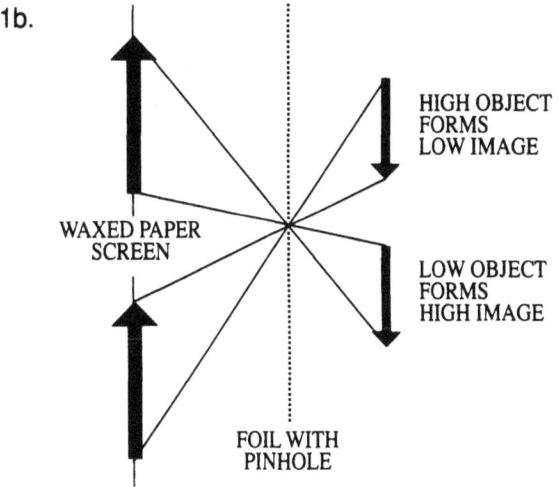

1a.

LARGE IMAGE   NEAR OBJECT

WAXED PAPER SCREEN   FOIL WITH PINHOLE

SMALL IMAGE   FAR OBJECT

1b.

HIGH OBJECT FORMS LOW IMAGE

WAXED PAPER SCREEN

LOW OBJECT FORMS HIGH IMAGE

FOIL WITH PINHOLE

2. Poke multiple pinholes into the foil. Each one forms a separate image on the waxed paper screen.

Materials

☐ A flashlight.
☐ A pinhole viewer from the previous activity.
☐ A straight pin.
☐ A sharpened pencil.

(TO) understand why shadows change size when objects are held close to a light source, but remain constant for distant sources.

SHADOW DISK O Light ( )

1. Tape white paper to an empty cereal box, forming a screen.

WHITE SCREEN

2. Cover the end of your flashlight with foil. Poke a hole in the middle, about as big as your pencil.

PENCIL-SIZED HOLE

3. Trace a circle around a battery onto an index card. Cut it out and attach a paper clip handle.

SHADOW DISK

4. Project shadows of this disk on your screen with the flashlight. Draw ray diagrams explaining how to make the shadow larger or smaller.

5. A distant light source, like the sun, produces light rays that strike the earth nearly parallel.

    a. Will the size of your disk's shadow change if you hold both it and the screen perpendicular to the sun's rays? Make a prediction.

    b. Test your prediction when the sun is shining.

RAYS FROM THE SUN

Save your screen and shadow disk to use again.

4

Answers / Notes

2. *Light from the flashlight is confined by foil to the relatively small area of a pencil hole to avoid multiple shadows. This phenomena will be explored in the next activity.*

5a. No. Because the sun's rays move in parallel, they may be intercepted at any position without blocking extra light.

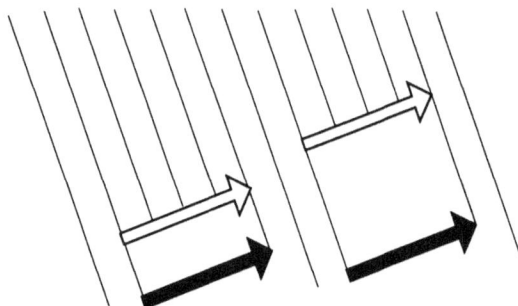

Another way to think about this is to remember that the disk is an extraordinary distance from the sun. An extra meter of distance nearer or farther is totally insignificant relative to this total distance.

5b. Students should test and evaluate their predictions. *Some may report, due to blurring around the edges, that the shadow looks smaller as you move the disk away from the screen. This effect will be studied in the next activity.*

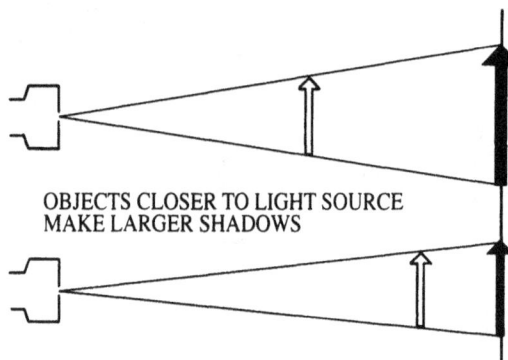

4.

OBJECTS CLOSER TO LIGHT SOURCE MAKE LARGER SHADOWS

Materials
☐ A sheet of white paper. Notebook paper is OK.
☐ Scissors.
☐ An empty cereal box.
☐ Tape. Use either masking or cellophane tape.
☐ A size-D battery, dead or alive. Only the flashlight needs live batteries. All other applications in this module are for non-energy uses.
☐ An index card.
☐ A paper clip.
☐ A flashlight.
☐ Aluminum foil.
☐ Direct sunlight. Students can complete all but step 4b on a cloudy day.

(TO) relate the clarity of a shadow to its distance from the light source and screen.

THE SHADOW IS FUZZY! ○ Light ( )

1. Mount a flashlight on 2 cans with rubber bands. Replace the foiled end with waxed paper.

WAXED PAPER

2. Mount your shadow disk on a battery with a rubber band. Align with your screen in a dim place to project a shadow.

DISK   SHADOW

3. Shadows are usually dark in the middle (the umbra) and lighter around the edges (the penumbra). Explain how to make a shadow that is…

    a. nearly all umbra.
    b. nearly all penumbra.
    c. an equal portion of umbra and penumbra.

PENUMBRA   UMBRA

4. Draw light rays from points x and y to explain why…

FLASHLIGHT   DISK   SCREEN

    a. an umbra and penumbra both form on the screen.
    b. the percentage of umbra increases as you move the disk nearer the screen.

Save your equipment for the next activity.

5

Answers / Notes

3a. Nearly all umbra: hold the disk far from the light source and close to the screen.
3b. Nearly all penumbra: hold the disk near to the light source and far from the screen.
3c. Equal portions: hold the disk at an intermediate distance.

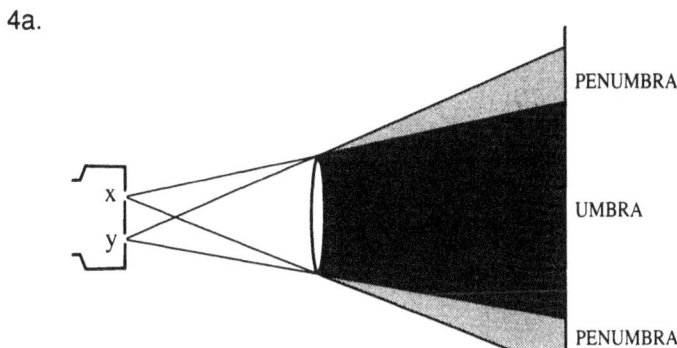

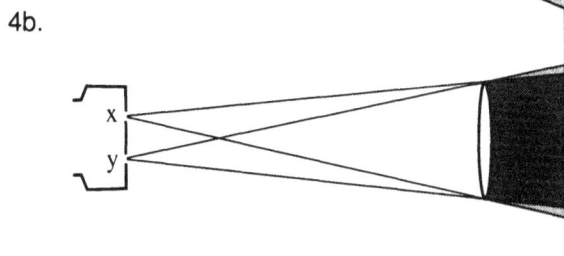

4a.

PENUMBRA

UMBRA

PENUMBRA

4b.

Materials

☐ A flashlight.
☐ Waxed paper.
☐ Rubber bands.
☐ Two cans. Use parts from the pinhole viewer constructed in activity 3.
☐ The index card shadow disk used previously.
☐ A size-D battery.
☐ The cereal-box screen from activity 4.
☐ A dim working area, away from direct lighting.

(TO) model a solar eclipse. To relate the completeness of an eclipse to the kind of shadow it casts on the earth.

SOLAR ECLIPSE   O   Light (   )

1. Punch a row of five holes across an index card, as far in as the punch will reach. Attach to a battery with a rubber band and paper clip.

2. Project a small umbra surrounded by a large penumbra on your screen. Position the index card so its holes span the complete shadow.

3. Now remove the screen to look at your "eclipse of the sun" through each hole. (Never do this with the real sun. It will damage your eyes.)

SUN: MOON: ECLIPSE:

o o o o o

4. Draw the parts of the "sun" and "moon" you see through each hole.

   a. Relate each part of this model to a real eclipse.

   b. Which drawing(s) show(s) that you are standing in the penumbra? In the umbra? Explain.

6

Answers / Notes

2. The middle hole should center in the umbra, while the holes to each side should extend well into the fading penumbra.

4.

4a. The flashlight disk covered with waxed paper represents the sun. The index card disk models the moon. The punched index card represents the various views that earth observers would see, depending on their positions within the umbra or penumbra.

4b. All but the middle drawing are views of the eclipse from within the partial shadow of the penumbra. Some light, from the rim of the "sun", enters each hole except the center one. This area of complete darkness is the umbra.

Materials

☐ An index card.
☐ A paper punch.
☐ A size-D battery, paper clip and rubber band.
☐ The experimental set-up from the previous activity.

(TO) discover that the angles of incidence and reflection are equal for light reflected by a plane mirror.

REFLECTION   O   Light ( )

1. Cut around the box containing the protractor. Fix it to your table with masking tape.

2. Tie a small loop in the middle of some thread, then tape its knot exactly where the normal meets the baseline. Tape the ends to pennies labeled *i* (for *incident ray*) and *r* (for *reflected ray*).

3. Lightly stick the back of a plane mirror to a battery with tape rolled sticky-side-out. Set the mirror directly on the baseline.

4. Slide the incoming ray, i, to any angle with the normal. Line up the outgoing reflected ray, r, with the reflection you see in the mirror.
    a. Compare the angle of incidence ($\angle$ i) with the angle of reflection ($\angle$ r).
    b. Does this relationship hold for all angles? Try and see.
    c. What is the value for $\angle$ i and $\angle$ r when you look at the pupil of your eye?
    d. Is it possible to reflect a ray for $\angle$ i = 90°?

5. Place a paper clip at $\angle$ i = 40°. Predict where you should hold a straw to see the image of this paper clip projected through it. Try it and see.

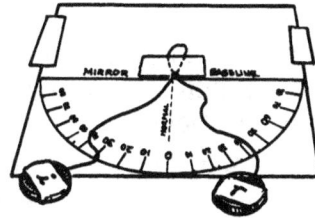

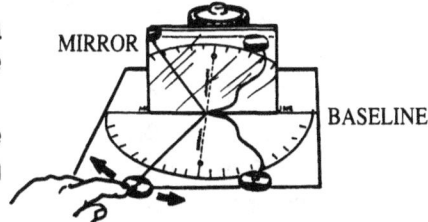

© 1991 by TOPS Learning Systems   7

Answers / Notes

3. The silver backing on some mirrors is fragile. Test one in advance to make sure that masking tape doesn't remove it. If it does, advise your class to tape their mirrors very lightly.

This reflecting surface should rest directly over the baseline. On most mirrors, it is painted onto the back of the mirror behind a thickness of glass; other mirrors have front reflecting surfaces.

4a. Within the limits of experimental error, $\angle$ i equals $\angle$ r.

4b. Yes. When $\angle$ i is larger or smaller, $\angle$ r is larger or smaller by an equal amount.

4c. Light from your eye pupil travels to the mirror along the normal — a path that is perpendicular to its surface — and then bounces straight back. Thus $\angle$ i = $\angle$ r = 0°.

4d. No. At $\angle$ i = 90°, the light ray travels parallel to the surface of the mirror, without being deflected in any way.

5. Hold the straw at $\angle$ r = 40°. The paper clip is easily seen through the straw at this position.

Materials

☐ A paper square containing a protractor. Photocopy this from the supplementary page at the back of this book.
☐ Scissors.
☐ Masking tape.
☐ Thread.
☐ Pennies.
☐ A battery.
☐ A small, rectangular, plane mirror.
☐ A paper clip.
☐ A straw.

(TO) discover that an object and its mirror image appear equidistant from the reflecting surface of a mirror.

LINE UP (1)　　　　　　　○　　　　　　　**Light (　)**

X　　　Y

1. Darken the middle line on notebook paper. Tape the *edge* of a mirror to a battery, then set its reflecting surface vertically over this baseline.

2. Cut a straw in half. Rubber band each half to other batteries so they stand perfectly straight.

3. Set *straw Y* at some distance behind the mirror. Position *straw X* somewhere in front of the mirror so its mirror image lines up perfectly with straw Y behind. The alignment is correct when you can move your head from side to side and still see only one straw behind the mirror.

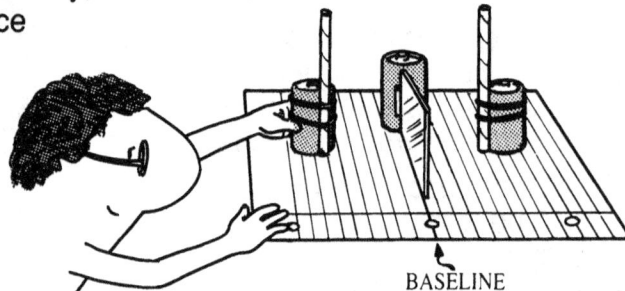

BASELINE

4. Count the lines from each straw to the baseline. What seems true?

5. Repeat the experiment with straw Y placed closer to the mirror. Do you get a similar result?

8

Answers / Notes

3. *All reflected light rays from straw X extend back to one unique point of convergence. If straw Y occupies this point, it lines up with the image of X, through every line of sight.*

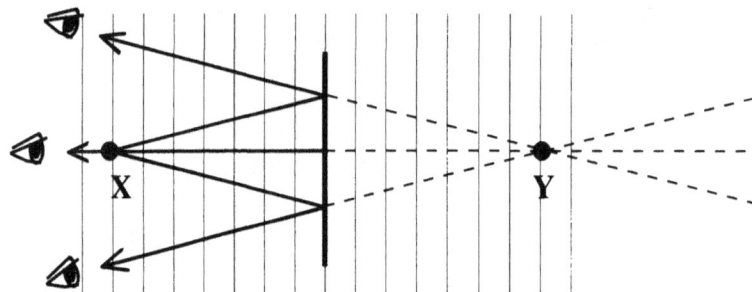

4. Within the limits of experimental error, the object straw and its reflected image appear equidistant from the reflecting surface of the mirror.

5. Yes. The object (straw X) and the apparent location of its reflected image (straw Y) both move closer to the mirror, remaining equidistant from the baseline.

Materials

☐ Notebook paper.
☐ Scissors.
☐ A plane mirror.
☐ Three batteries.
☐ Masking tape.
☐ A straw.
☐ Rubber bands.

(TO) predict the position of images behind a mirror by drawing rays and measuring angles. To confirm the accuracy of these drawings by checking their reflections in a plane mirror.

LINE UP (2)   O   Light (   )

1. Fold notebook paper in half lengthwise. Mark the crease to form a baseline. Draw an x below, exactly on a notebook line. This represents an object and a mirror.

2. Draw an incident light ray from this x along the normal. Draw two more rays, each meeting the mirror at intersecting lines.

3. Show how these rays reflect so ∠i = ∠r. Accurately draw and label all angles.

4. Extend the reflected rays behind the mirror using dashed lines. Where they cross locates the mirror image of x.

5. Stand a mirror on the baseline. Is the reflected x positioned where you predicted it should be? Explain.

6. Repeat this experiment with a triangle, drawing the fewest possible rays to locate its image behind the baseline. Again, check your accuracy with a real mirror.

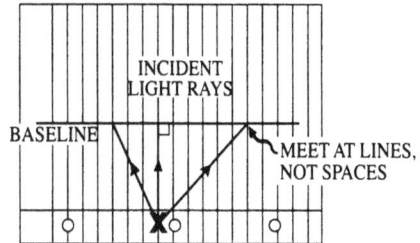

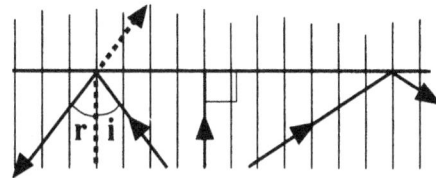

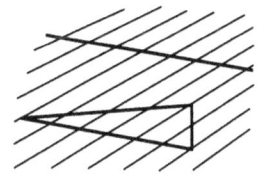

INCIDENT LIGHT RAYS
BASELINE
MEET AT LINES, NOT SPACES

9

Answers / Notes

1-4.

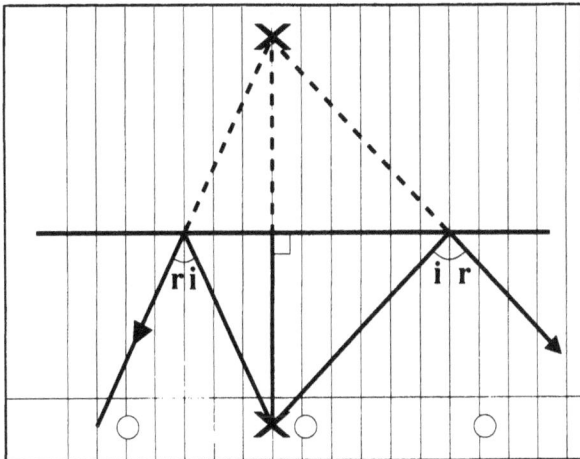

5. Yes. The reflection of the x and its plotted position appear superimposed as you move your head up and down over the top edge of the mirror.

6. *If the triangle and its rays are well chosen, this construction can be greatly simplified. Notice how six different lines of notebook paper all serve as normals.*

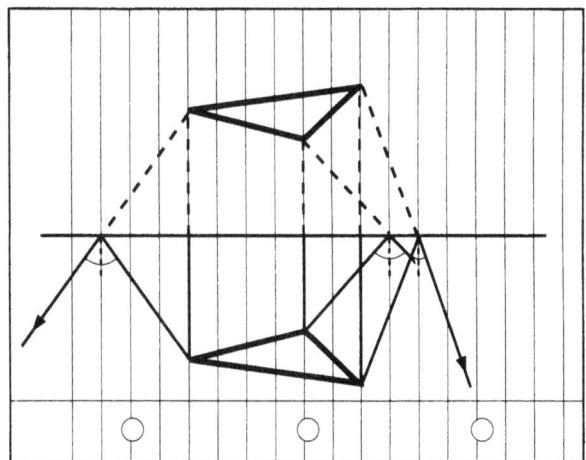

Materials

☐ Notebook paper.
☐ A protractor. If students use the paper version photocopied from activity 7, they should first cut off its surrounding square background.
☐ An index card or other straight edge.
☐ A plane mirror.

(TO) diagram where virtual images are formed behind plane mirrors, using the principle that the angle of incidence equals the angle of reflection.

UP OR DOWN? O Light ( )

1. The image behind a plane mirror is said to be *virtual*. Light doesn't really come from the candle behind the mirror, but it looks virtually like it does.

VIRTUAL IMAGE

a. Does the eye interpret light rays as bouncing off the mirror? Why are dashed lines used in this diagram?

b. Set a battery on your head and look at its reflection in a mirror. Draw a ray diagram to show the relative positions of your eye, the battery and its virtual image.

2. Position 2 mirrors to observe what is behind you. Flip the *top* mirror to reflect what is ahead.

LOOKING BEHIND   LOOKING AHEAD

a. How is the scene behind you different from the one ahead?

b. Account for these differences using ray diagrams.

10

Answers / Notes

1a. No. The eye interprets the position of an object by the direction its light enters the eye to strike the retina. The candle seems to lie behind the mirror because its light appears to come from that location. Dashed lines are used because these do not represent real light rays, rather virtual rays that the eye interprets as real.

1b.

VIRTUAL IMAGE

2a-b. *These drawings can be greatly simplified by allowing students to omit all virtual rays and images.*

Objects reflected from behind are *inverted*.

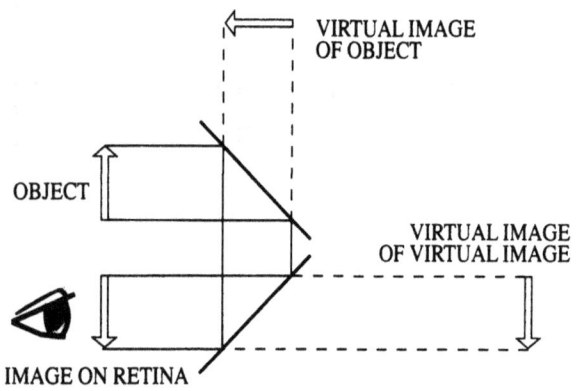

VIRTUAL IMAGE OF OBJECT

OBJECT

VIRTUAL IMAGE OF VIRTUAL IMAGE

IMAGE ON RETINA

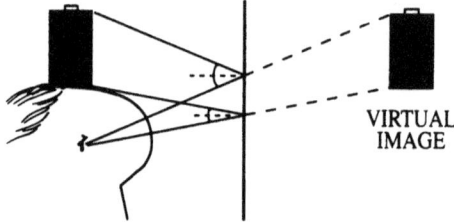

Objects reflected from in front are *erect*.

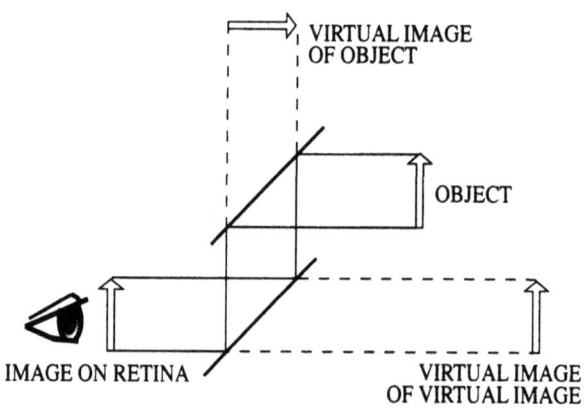

VIRTUAL IMAGE OF OBJECT

OBJECT

IMAGE ON RETINA

VIRTUAL IMAGE OF VIRTUAL IMAGE

Materials

☐ A battery.
☐ Two large hand mirrors. Smaller rectangular mirrors will also work, but are more difficult to align.

(TO) create an illusion with mirrors. To explain the illusion using principles of light reflection.

FUNNY FACES   O   Light ( )

1. Slide 2 mirrors together so their middle junction runs through the image of both your eyes.

2. Put a penny under the edge of each mirror to give your face four eyes.

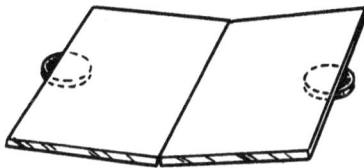

3. Slide the pennies under the crack in the middle of the mirrors to give yourself no eyes!

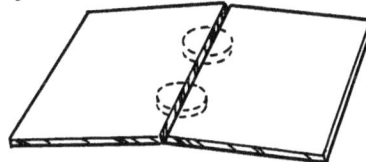

4. Use diagrams to explain the illusions in steps 2 and 3.

11

Answers / Notes

4. When these mirrors are tilted slightly down in the middle, double images for each eye are reflected toward the observer, creating an illusion of 4 eyes.

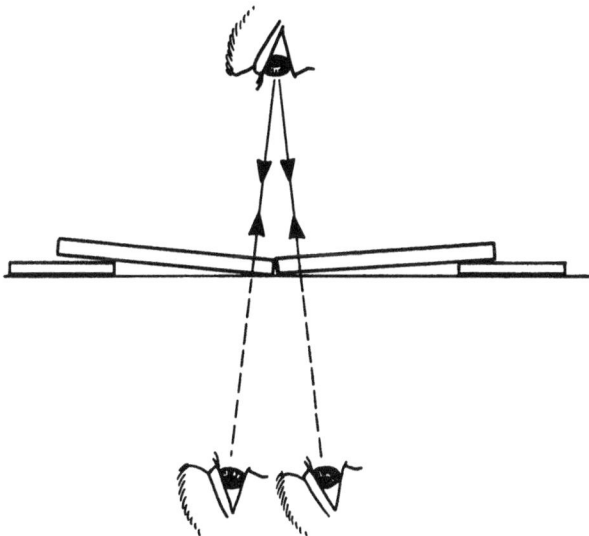

When the mirrors are tilted slightly up in the middle, the two images for each eye are reflected away from the observer, creating an illusion of no eyes.

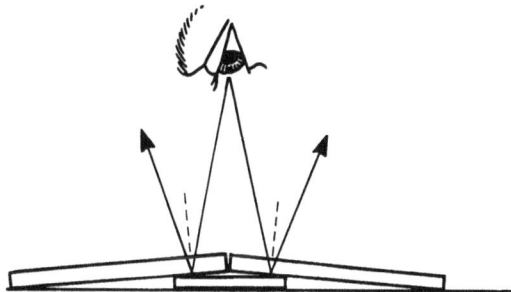

Materials

☐ Small unframed rectangular mirrors. The reflecting surface should extend to the edges of each mirror. If not enough mirrors are available, substitute microscope slides with black construction paper as a background.
☐ Two pennies.

(TO) use the transmitting and reflecting properties of glass to create visual illusions.

VIRTUAL REALITY ◯ Light ( )

1. Cover the head of your flashlight with foil. Poke a pencil hole in the center.

2. Attach the long edge of a glass microscope slide to a battery with masking tape.

3. Fill a jar with water. Use all of these objects to make a spot of light appear as if it were virtually floating under water.

4. Tell how you created this illusion. Explain how it works.

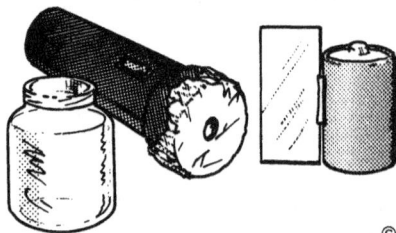

5. Put a penny on white paper. Hold a glass slide on edge just behind it.

6. Tip the glass slide forward so you can see the image of the penny reflected in the slide.

7. How should you move the slide to make a penny image first appear, then fade and disappear.

8. Light both reflects off glass and passes through it in amounts that depend on ∠i. Use this concept to explain why the penny fades.

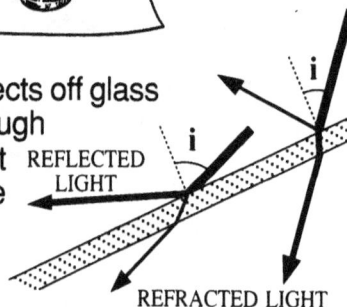

REFLECTED LIGHT

REFRACTED LIGHT

© 1991 by TOPS Learning Systems

12

Answers / Notes

4. To create this illusion, position the flashlight in front of the microscope slide and the glass of water an equal distance behind it. Light from the jar of water passes *through* the slide into your eyes. Light from the flashlight *reflects off* the slide into your eyes, creating a virtual reflection at this same location.

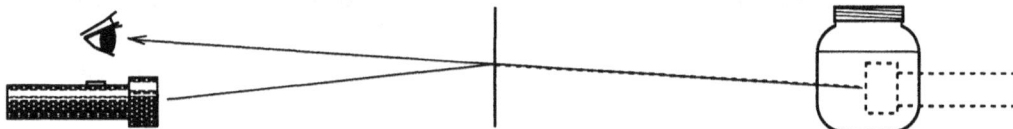

7. Tip the slide forward to make the penny image appear. Tip it back to make the image fade and disappear.

8. When the glass slide is tipped forward and the light from the penny strikes it at a greater ∠i, more is reflected toward the observer's eye, producing a sharp, clear image.

As the glass slide is tipped up, ∠i decreases and the image fades. More of the penny's light now passes through the glass; less is reflected toward the observer's eye.

Finally the glass slide is tipped so far back that no reflected light is directed toward the observer's eye. This causes the image to disappear completely.

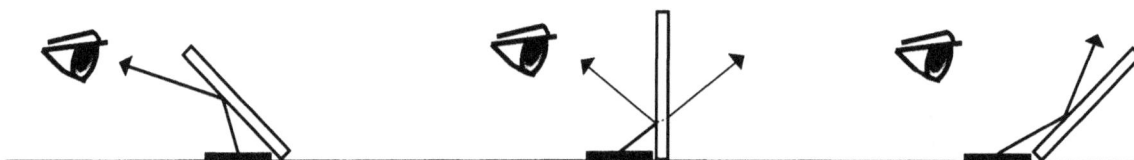

Materials

☐ A flashlight and piece of foil. You can create a more dramatic illusion of fire under water by substituting a lighted candle. If you do, its total height, including the flame, should not exceed the length of the microscope slide. A half birthday candle has the correct height. Or you might use longer candles and substitute larger panes of glass.

☐ A battery.
☐ Masking tape.
☐ A microscope slide.
☐ A jar or glass of water.
☐ A penny.

(TO) examine how water waves in a pan model the behavior of light waves.

LIGHT AS WAVES   O   Light (   )

1. Make a ripple tank: add enough water to a rectangular cake pan to *just* cover a straw.

2. Drip water from an eyedropper high above the center of the calm water.
   a. How do water waves travel outward from the drop?
   b. Do water waves reflect off all sides of the pan?
   c. Imagine striking a match in a totally dark room to see your surroundings. How do water waves in a pan model this event?

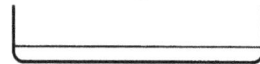

3. Trim the straw (if necessary) to easily fit inside the width of the tank. Add a masking tape handle in the middle of the straw.
   a. Make waves by moving the straw up and down at one end of the tank. Do these waves reflect? Does $\angle i = \angle r$?
   b. Now make waves by moving the straw up and down at the *corner* of the tank (diagonally). Do these waves reflect? Does $\angle i = \angle r$?

(Save your taped straw.)

© 1991 by TOPS Learning Systems   13

Answers / Notes

2a. Water waves travel out in all directions within the plane defined by the water's surface.

2b. Yes. The waves first reflect off the sides of the pan, then off the more distant ends.

2c. Just as water waves travel out from the drop in all directions and reflect off the sides of the container, light travels out from the flame in all directions and reflects off the walls of the room.

3a. Yes. The water waves travel to the back of the pan along the normal and then are reflected straight back. Thus $\angle i = \angle r = 0°$.

3b. Yes. The water waves come toward the side of the pan at an angle and then reflect away at the same angle. This holds true for large and small angles of incidence.

These reflected waves may be difficult to see at first. They are best viewed with $\angle i = 45°$.

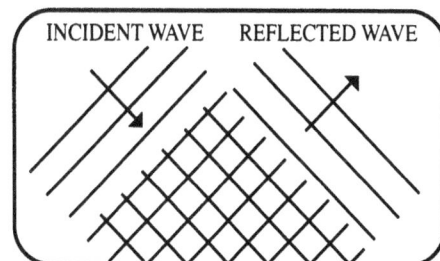

INCIDENT WAVE   REFLECTED WAVE

Materials

☐ A rectangular cake pan about 9 by 12 inches (20 x 30 cm) or larger. Rest the pan on a flat surface to make sure it has not been too warped by heat.
☐ A flat table or other surface.
☐ A source of water.
☐ A straw.
☐ An eyedropper.
☐ Masking tape.
☐ Scissors.

(TO) bend water waves in a ripple tank by slowing them against a shallow bottom. To relate this behavior to the bending of light beams.

REFRACTION (1)   O   Light (   )

1. Set up your ripple tank on a flat surface as before, with enough water to just cover your wave-generating straw.

2. Now tip the pan by sliding a pencil under one edge. Generate *wave fronts* as before and record your observations.

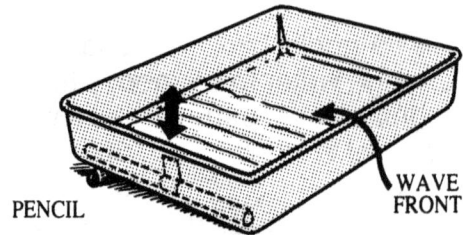

PENCIL   WAVE FRONT

3. Waves slow down in shallow water due to friction against the bottom.
   a. Explain how this bends an advancing wave front.
   b. Bend the wave front in the opposite direction. Explain how you did this.

4. Rubber band 2 batteries to a flashlight. Rest this at an angle on an open can so it shines a spot of light against the bottom of the emptied pan.

WAVE FRONTS

5. Mark the near edge of this light spot with a paper clip. How does this spot shift as you add more and more water to the pan?

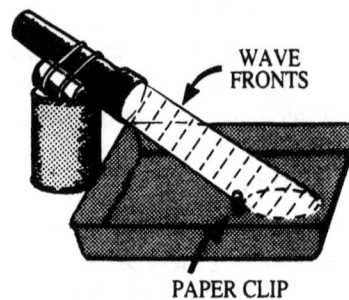

6. Light waves "drag" more slowly through water than air. How does this fact explain what you see?

PAPER CLIP

© 1991 by TOPS Learning Systems   14

Answers / Notes

2. The waves bend toward the shallow side of the pan.

3a. The shallow end of the advancing front moves slower than the deep end because of greater friction with the bottom. This allows the part of wave over the deep end to move a greater distance than the part over the shallow end, thus turning the wave front into the shallow side.

SHALLOW: slower moving

DEEPER: faster moving

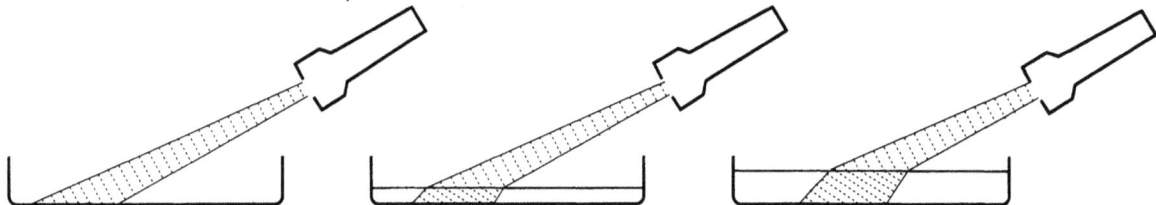

3b. To bend the waves the other way, reverse the shallow and deep sides by putting the pencil under the opposite side of the pan.

5. As water fills the pan, the light spot shifts towards the paper clip and flashlight.

6. Without water, light from the flashlight strikes the bottom of the pan along a straight path. With water, the light bends at the air-water interface. The waves of light turn towards the water, in the direction of increased drag, just as water waves turned in the pan.

Materials

☐ The cake pan and straw from the previous activity.
☐ Two batteries, a can and a flashlight. Size D batteries nest perfectly, with flashlight attached, into a 15 ounce vegetable can.
☐ Rubber bands.
☐ A paper clip.

(TO) diagram how an incident light ray is refracted through glass. To account for its direction of bending.

REFRACTION (2) O Light ( )

1. Place 4 glass slides on edge. Draw an outline around them on paper.

2. Draw and label a line to represent an incoming incident light ray as shown.

3. Sight through the glass with an index card where this incident ray emerges on the other side. Draw this line of sight and label it the refracted (bent) light ray.

4 SLIDES

INCIDENT RAY

REFRACTED RAY

OUTLINE

4. Remove the slides. Join the incident and refracted light rays with another straight line. Label the points of refraction.

5. Think of your pencil as an advancing wave front. Does light travel faster or slower as it passes through glass? Explain.

6. Which property of light do you think best explains refraction — its particle nature or wave nature?

© 1991 by TOPS Learning Systems

15

Answers / Notes

1-4.

incident ray

REFRACTION

REFRACTION

reflected ray

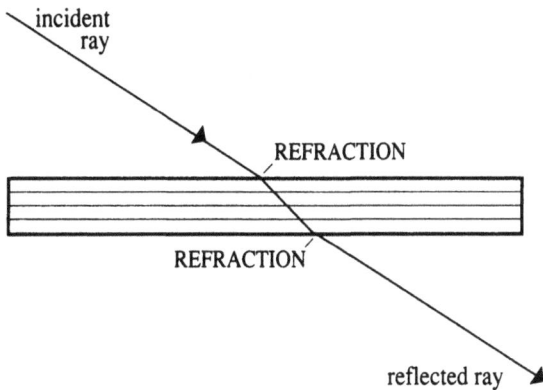

5. Light travels more slowly through glass than through air: as the advancing wave front passes from air to glass, one side of the wave front enters the glass before the other. The side that enters first moves slower, since the ray bends in that direction. Then, as the ray emerges from glass back into air, the side of the wave front that emerges first moves faster, bending the ray the opposite way.

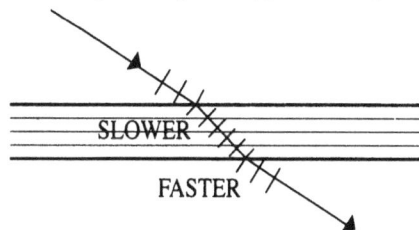

SLOWER

FASTER

6. *This question asks students to express an opinion. Answers, therefore, should not be judged as right or wrong, but as reasonable or unreasonable. (We think the wave nature of light best explains refraction. It's easy to imagine part of a wave front dragging in a denser medium, while the remainder of the front moves faster, swinging the front into a new direction of travel. It's not so easy to imagine how particles might change direction as they travel through different mediums, unless, like a golf ball, they acquire some sort of spin.)*

Materials

☐ Four glass slides.
☐ An index card.

(TO) predict how the direction and amount of bending of refracted light varies with a changing angle of incidence. To verify these predictions by experiment.

REFRACTION (3)　　　　　○　　　　　Light (　　)

1. Think of your pencil as an advancing wave front. Describe how light should refract in each group of 4 slides. Explain your reasoning.

a.　　　　　　　　　b.　　　　　　　　　c.

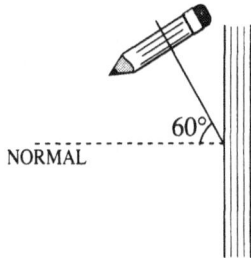　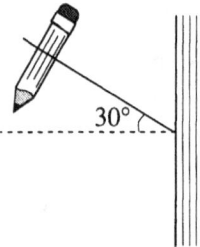　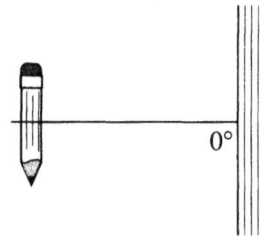

2. Test each experiment by drawing incident light rays at the correct angles and sighting through the slides as before. Were your predictions correct?

3. Pivot the 4 slides, as a group, over lines of notebook paper like this. Do the lines appear to move as you predicted above? Explain.

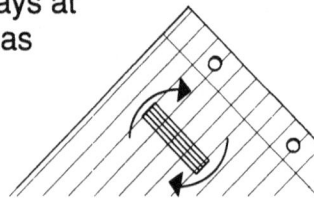

© 1991 by TOPS Learning Systems　　　　　16

Answers / Notes

1a. The light ray will bend into the glass, as the top of the advancing front is slowed by the glass before the bottom. Similarly, as this advancing front emerges back into air, it will bend in the opposite direction as the top speeds up before the bottom.

1b. The light ray will again bend toward the glass, both entering and exiting the medium. This time, the amount of bending will be less. Due to a decreased angle of incidence, each side of the advancing wave front will be independently outside or inside the glass over a shorter period of time.

1c. There will be no bending. All parts of the wave front enter and exit the glass together, simultaneously slowing down and speeding up.

2a.
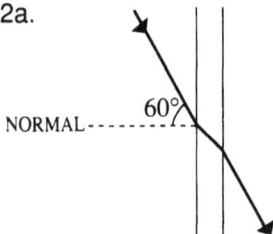

2b.
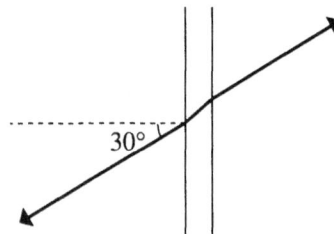

2c.
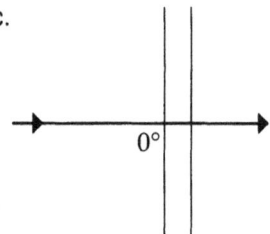

Students should comment on the accuracy of each of their predictions.

3. Yes. The lines behind the slides appear to shift relative to the lines in front of the slides as they are turned through a range of incident angles. This shift is most pronounced at high angles of incidence, then fades to no refraction at all as the angle of incidence approaches 0°.

Materials

☐ A protractor.
☐ Four microscope slides.
☐ An index card.

(TO) understand how refracted light changes the apparent position of an object in water when viewed from above its surface.

WHERE'S THE PENNY?　　　O　　　　　Light ( )

1. Place a penny at the bottom of a shallow can. Position your eye so the penny is just out of sight.

    a. Without changing your line of sight, add water to the can. What do you notice?

    b. Explain your observations with a diagram.

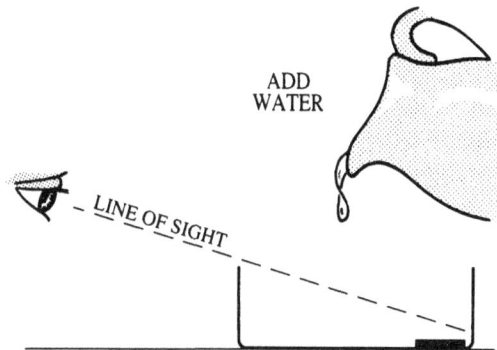

2. With the can full of water, try to "spear" the penny with a straw.

    a. If you aim directly at the penny, can you hit it?

    b. Explain your observations with a diagram.

ADD WATER

LINE OF SIGHT

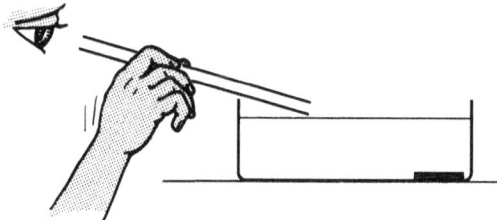

17

Answers / Notes

1a. The image of the penny appears to rise into view as you add water.

1b. The light refracts at the water's surface and bends over the top of the can into your eyes.

2a. No. If you aim directly at the image of the penny, the straw lands too high.

2b. The light refracts at the water's surface, creating a virtual image that is higher than the actual penny.

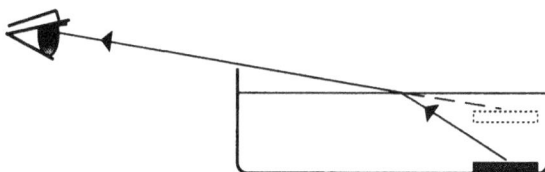

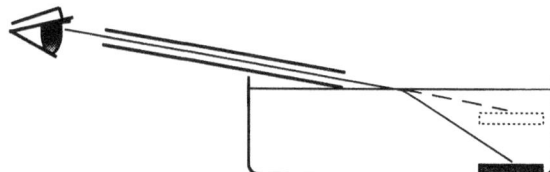

Materials

☐ A tuna fish can or similar opaque container.
☐ A penny.
☐ A straw.
☐ Water.

(TO) create concave, planar and convex water surfaces in a test tube. To understand why these shapes produce images of different sizes.

WATER LENS ○ Light ( )

1. Wrap a rubber band twice across the mouth of a small jar.

2. Push 2 small test tubes into the jar between the bands.

3. Unbend 2 paper clips in the middle, then pull out the large end just a little so they wedge inside each tube, about 1/3 down from the rim.

4. Fill both tubes with water; the jar half full.

5. Use an eyedropper to transfer water between the tubes and jar. Tell how to make these different surface shapes in the tubes:

PUSH PAPER CLIP 1/3 DOWN

DOUBLED RUBBER BAND

CONCAVE   PLANAR   CONVEX

6. How does each surface shape affect the size of the paper clip image as you look into the top of the test tube?

7. Examine a hand lens: How does its shape affect image size?

© 1991 by TOPS Learning Systems   18

Answers / Notes

5. Concave surface: Keep the surface of the water below the rim of the tube.

    Planar surface: Add just enough water to bring its surface even with the rim.

    Convex surface: Add extra water to hump the surface of the water over the rim.

6. A concave surface makes the paper clip look smaller.

    A planar surface doesn't change the apparent size of the paper clip.

    A convex surface makes the paper clip look larger.

7. A hand lens is convex on both sides (double convex). Like the convex water surface, it enlarges images.

Materials

☐ A baby food jar or small beaker, plus a rubber band large enough to wrap twice around it.
☐ Two small test tubes. Use the narrowest possible diameter to produce the widest variation in image size. The second tube is added as a control to facilitate image size comparisons.
☐ Two paper clips.
☐ A source of water.
☐ An eyedropper.
☐ A hand lens. Ones with straight, rigid handles work best for most activities in this module. These may be ordered through scientific supply outlets, or from our catalog, *TOPS Ideas*.

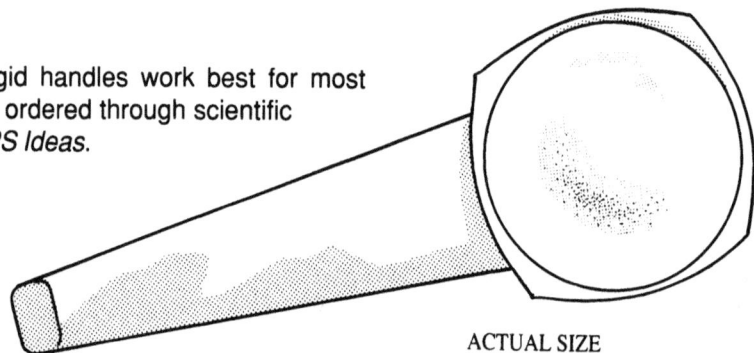

ACTUAL SIZE

(TO) interrupt projected light rays with a variety of objects. To interpret the various reflections and refractions that result.

RAY PLAY ○ Light (   )

1. Cover the end of a flashlight with aluminum foil. Cut a straight, narrow slit (about ⅓ across) that reaches to the rim but stops short of the center.

FOIL COVER

SLIT

2. Lay the flashlight, so that light shining through the slit casts a bright, narrow beam across white paper. (A dark work area helps you to see this most clearly.)

3. Interrupt this beam (near the slit) with each material below, experimenting with different angles. Report your findings in words and pictures.

    a. A small unframed mirror with straight sides.

    b. A sandwich of four microscope slides.

    c. A transparent pill vial (with and without water.)

© 2000 by TOPS Learning Systems   19

Answers / Notes

1. *Unless your students can handle knives or razor blades safely, cut this slit for them. Make it a little shorter than the radius of the flashlight window. Slits cut up to, or across the center of the light bulb tend to cast irregular light beams, because of wide variations in intensity.*

2. *A narrow, well-defined beam should project a minimum of 6 cm across the paper.*

3a. Mirror: The ray reflects off the surface of the mirror. As the angle of incidence changes, so does the angle of reflection: both angles are equal at all times.

3b. Microscope Slides: For high incident angles, most of the ray reflects off the glass while a smaller portion of the light refracts through the glass. As this incident angle decreases towards 0°, a smaller portion of the ray reflects while a larger portion of light refracts on through. And the amount of bending continues to decrease.

3c. Pill Vial: Without water, the ray passes through the empty container relatively unaffected, except along the edge where some refraction and reflection take place. With water in the bottom the beam is dramatically refracted.

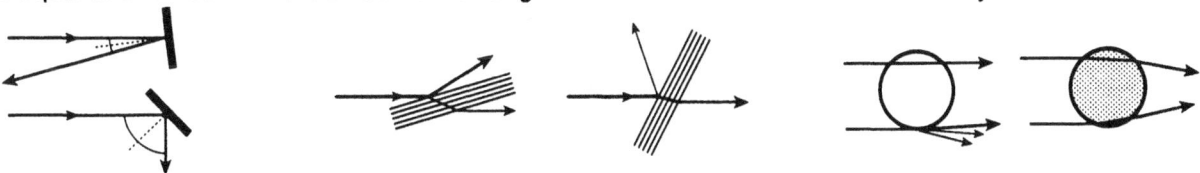

Materials

☐ A flashlight with fresh D cells.

☐ Aluminum foil. Precut squares to about 8 cm (3 inch).

☐ A sharp knife or razor blade to cut the narrow slit. Even a straight pin will cut foil, but leaves a ragged edge.

☐ A white sheet of scratch paper. Notebook paper is OK

☐ A work area that may be somewhat darkened.

☐ A small unframed mirror with straight sides that reflects light at the outside edge. Round or framed mirrors will not work.

☐ Four microscope slides. Prisms also reflect and refract in interesting ways.

☐ A transparent pill vial or equivalent. Any clear narrow plastic drinking cup or glass beaker with a flat bottom is also suitable. Containers with rounded bottoms (baby food jars, for example) will not work.

☐ A small portion of water.

(TO) identify colors in the visible spectrum. To recognize that these colors are the refracted components of white light.

THE COLOR SPECTRUM   O   Light (   )

1. Set a small jar on the bottom of a can. Rest both off center on a white plate.

2. Direct a source of light across the top of the jar at a sharp angle so the can's shadow falls across the plate.

3. Fill the jar with water until it brims above the lip. The rounded edge of water should refract a *spectrum* of color onto the shadowed plate.

4. List all the colors in this spectrum, starting farthest from the jar and working in.

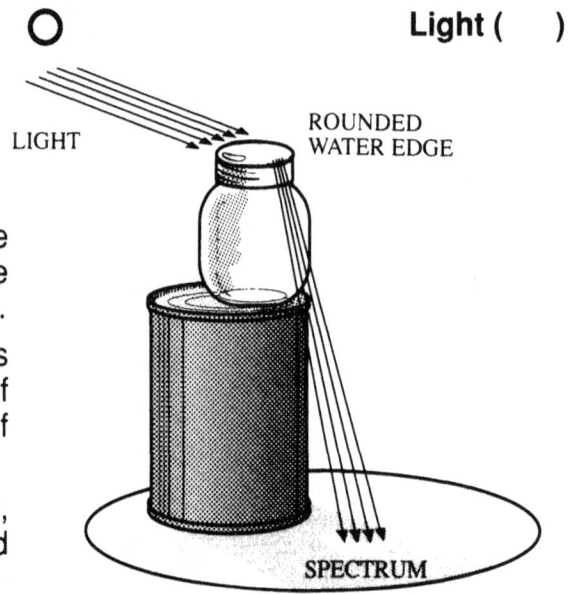

5. Red has the longest wavelength of all the colors; violet the shortest. How does wavelength affect a color's tendency to refract?

6. Is white light a pure color?

LIGHT

ROUNDED
WATER EDGE

SPECTRUM

20

Introduction

Light travels through space like a wave, partly electric and partly magnetic. It fits between radiated heat (infrared) and ultraviolet light on the electromagnetic spectrum.

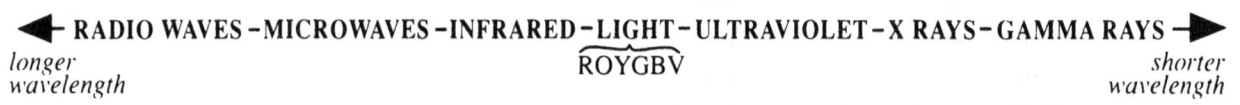

◀— RADIO WAVES – MICROWAVES – INFRARED – LIGHT – ULTRAVIOLET – X RAYS – GAMMA RAYS —▶

longer wavelength   ROYGBV   *shorter wavelength*

The eye responds to different frequencies in this visible part of the spectrum by seeing different colors. Red has the longest wavelength; violet the shortest.

Answers / Notes

4. red, orange, yellow, green, blue, violet.

5. Shorter wavelengths of light bend through sharper angles than longer wavelengths. Thus, violet is closest to the jar, red farthest away.

6. No. White light is a blend of all colors in the visible spectrum. Only when this light is sharply refracted are its various wavelengths isolated and perceived as different colors.

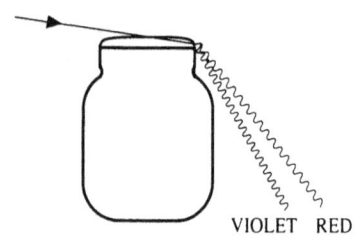

VIOLET   RED

Materials

☐ A small jar and can. A baby food jar coordinates well with a 15 ounce vegetable can.

☐ A white plate — paper or ceramic. White paper also serves, but won't contain spilled water.

☐ A source of water and eyedropper.

☐ A strong light source. The sun is best but a flashlight will also serve. If you are using artificial light, test it in advance to insure that all colors in the spectrum are present. Overhead sunlight may be redirected at a high angle of incidence by reflecting it with a mirror.

☐ Prisms are nice, but totally optional. Use at least one per lab group. Instruct students to first experiment with their prisms using a strong light source and white paper, then answer questions 4-6 on this task card. If you have just one prism, consider displaying its beautiful spectrum in your darkened classroom.

(TO) mix colors by subtracting them from white light and by adding them to a white background. To distinguish between color addition and subtraction.

ADDITION AND SUBTRACTION ○   **Light (   )**

1. Most colored objects absorb and transmit a range of colors. Blue and yellow cellophane, for example, absorb and transmit these colors:

    a. Predict the color you get by passing white light through *both* blue and yellow cellophane. Explain.

    b. Predict the color you get by *mixing* blue and yellow light on white paper. Explain.

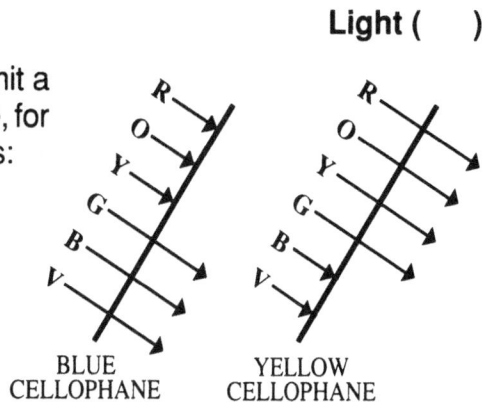

BLUE CELLOPHANE   YELLOW CELLOPHANE

2. Rubber band some yellow cellophane around a mirror, letting it stick out at one end. Do a second mirror in blue. Experiment with sunlight to test your predictions.

SUNLIGHT   COLOR?   COLOR?

3. Yellow + blue yields two different colors. How can this be?

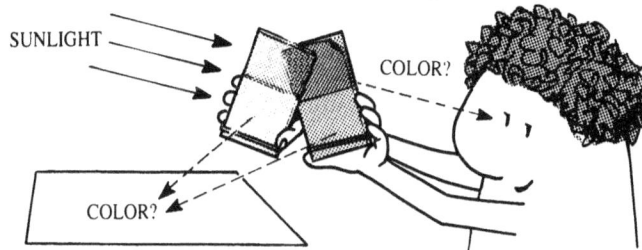

21

Introduction

Sunlight, composed of all colors in the spectrum, shines onto white, black and blue squares as shown.

a. What colors does each square reflect? *White reflects nearly all colors; black reflects almost no color; blue reflects blue (short wavelengths in the spectrum.)*

b. What colors does each square absorb? *White absorbs almost no color; black absorbs nearly all colors; blue absorbs colors that are not near blue (long wavelengths in the spectrum.)*

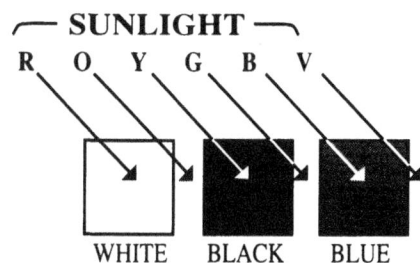

SUNLIGHT
R   O   Y   G   B   V

WHITE   BLACK   BLUE

Answers / Notes

1a. Green. This is the only color transmitted by both pieces of colored cellophane.

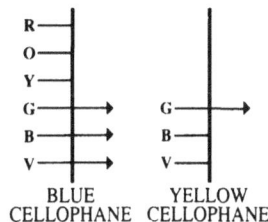

BLUE CELLOPHANE   YELLOW CELLOPHANE

1b. White. Blue cellophane transmits the shorter wavelength colors, while yellow cellophane transmits the longer. Where the colored lights overlap, full spectrum white is present.

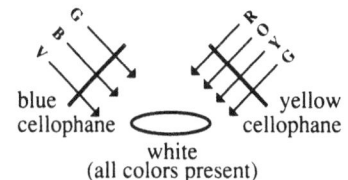

blue cellophane   yellow cellophane
white (all colors present)

2. White light that is filtered by overlapping blue and yellow cellophane looks green. But if both colored lights are reflected onto white paper by the mirrors, it looks white where they overlap. *(If the overlapping color trends toward blue or yellow, add more or fewer layers of cellophane until neither color dominates.)*

3. Green is the result of *subtracting* or filtering out the blue and yellow parts of the spectrum. White is the result of *adding* the blue and yellow parts of the spectrum together.

Materials

☐ Blue and yellow cellophane.
☐ Scissors.
☐ Two mirrors.

☐ Rubber bands.
☐ Direct sunlight. Strong full-spectrum light from a slide projector will also serve. Cellophane-covered flashlights give less satisfactory results unless your room can be made quite dark.

(TO) observe how a hand lens refracts incoming parallel rays to a focal point. To measure its focal length.

FOCAL LENGTH (1) ◯ Light ( )

1. Crease 2 sheets of notebook paper on the red margin lines. Wrap them around 2 books of equal size so each fold defines the top edge of each spine. Hold with rubber bands.

2. Press the book spines together, then sink exactly half of a hand lens between them. Keep the spines parallel.

3. Darken the edge of an index card with the side of your pencil. Align this dark edge with each blue line you see through the lens, then draw along the card.

   a. Your lens *converges* parallel lines to a *focal point*. What does this mean?

   b. Measure the *focal length* — the distance from the focal point to the center of the lens.

   c. Why do parallel "light rays" refract more towards the edges of the lens and less towards the center?

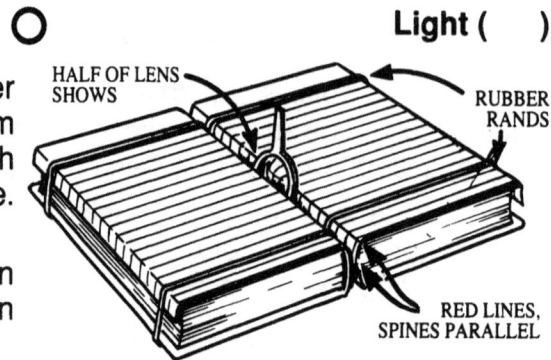

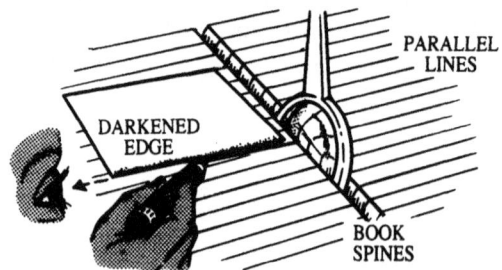

© 1991 by TOPS Learning Systems   22

Answers / Notes

2. *The book spines should be kept parallel. The blue lines on each piece of notebook paper don't necessarily need to line up here, but they will in later activities.*

3.

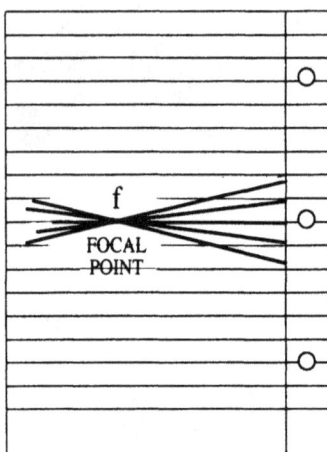

3a. The lines on the far side of the lens appear to come together (converge) to a point (focal point) when refracted by the lens.

3b. *Students should measure beyond the red margin line to the center of the lens. (The hand lenses sold by TOPS have a focal length of 8.2 cm. Other lenses, of course, may have different focal lengths.)*

3c. Parallel light rays toward the edges of the lens refract more sharply because they strike its surface at greater angles of incidence.

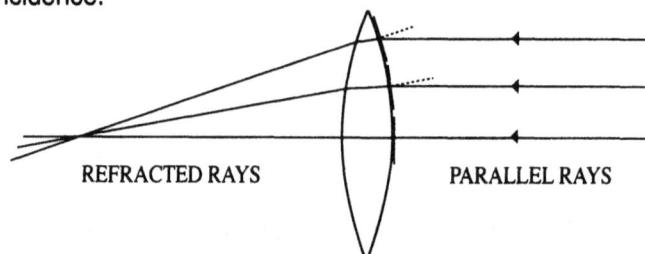

Materials

☐ Two identical textbooks with enough cover area to accommodate sheets of notebook paper. A moderate paper overhang is OK.

☐ Rubber bands.

☐ Identical hand lenses. If you are using an assortment of lenses, have each student or lab group identify their own with a label to use throughout this module.

☐ An index card.

☐ A metric ruler. Photocopy the ruler on the supplementary page at the back of this module, or supply your own.

(TO) observe how light from a distant source focuses through a lens at the focal point. To observe a cone of focusing light rays in a jar of cloudy water.

FOCAL LENGTH (2)　　○　　　　　　Light (　)

1. Bend a paper clip at a right angle. Tape the narrower end to the handle of your lens like this:

INDEX CARD SCREEN
STRAW
f

2. Push the wider end into a straw. Cut it to the focal length you determined before.

3. Hold your lens 1 focal length from an index card while facing it toward a distant scene outside a window. What do you see? (If the image isn't sharpest at the end of the straw, revise your accepted focal length.)

LIGHT

4. Make a jar of water slightly cloudy with a tiny pinch of powdered milk. Rubber band your straw (with lens) to the side of the jar, then reflect sunlight from a mirror or distant bright light down into the water. Block out excess light with a dark cloth or towel.

CLOUDY WATER
DARK CLOTH
RUBBER BAND

   a. Sketch the light cone you observe.
   b. Why does it focus beyond one focal length?

© 1991 by TOPS Learning Systems　　　　　　　23

Introduction

Show by diagram how light rays from more distant objects enter a lens more nearly parallel than rays from closer objects.

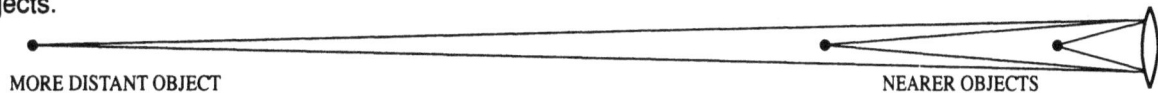

MORE DISTANT OBJECT　　　　　　　　　　NEARER OBJECTS

Where would you expect a lens to focus light rays from a distant source? *(Very near the focal point; the same place where parallel lines on notebook paper appeared to focus in the last activity.)*

Answers / Notes

2. Ideally, the straw should be slightly shorter than the focal length, since it meets the lens at a point slightly in front of its central axis.

3. The scene outside the window appears inverted and reduced on the index card screen.

If this image is not in focus at the end of the straw, then the card should be adjusted nearer or farther from the lens, as necessary, until it is. Once the image is sharp and clear, the corresponding focal length shold be remeasured and recorded.

4a-b. The cone extends beyond one focal length because it is refracted, bent outward, as it passes from air into water.

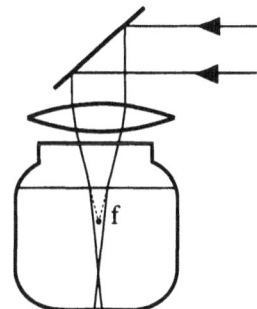

f

Materials

☐ A hand lens.
☐ A paper clip.
☐ Masking tape.
☐ A straw.
☐ Scissors.
☐ A metric ruler.
☐ A dark colored cloth.

☐ A window facing to the outside. A "scene" from an overhead projector or any bright light held at least 10 feet away also qualifies as a distant, "infinite" light source. The light should not be round, however, since students may not detect that its image is inverted.
☐ Direct sunlight for step 4. If the day is cloudy, substitute a bright distant light or a flashlight held high overhead in a relatively dark room.
☐ A baby food jar or beaker of water.
☐ Powdered milk. Fresh milk will also serve.

(TO) classify incident light rays into 5 possible groups and examine how they are characteristically refracted by a hand lens.

RAY RULES　　　　　　　　○　　　　　　　**Light (　)**

1. Set your lens between books covered with notebook paper, as before. Draw the *principal axis* — the normal that extends through its exact center.

2. Align all lines on each paper. Measure and mark the two focal points on the principal axis.

3. Draw 2 long straight pen lines on plastic wrap, then cut around each one.

ALL LINES SHOULD BE ALIGNED

PRINCIPAL AXIS

PEN LINE

PLASTIC WRAP

4. Line up both plastic "light rays" through the lens to complete each statement below. Illustrate each answer with a diagram that contains at least 2 rays. An incident light ray that:
    a. moves *parallel to the principal axis*, refracts…
    b. crosses the principal axis through its *focal point*, refracts along a path that…
    c. crosses the principal axis through the *center of the lens*…
    d. crosses the principal axis *inside the focal point*, refracts along a path that…
    e. crosses the principal axis *beyond the focal point*, refracts along a path that…

© 1991 by TOPS Learning Systems　　　　24

Introduction

Write these vocabulary words on your blackboard. Illustrate their meaning with arrows moving inward and outward.

CONVERGE　　DIVERGE

Answers / Notes

4. An incident light ray that…

a. moves *parallel to the principal axis* refracts <u>through the focal point</u>.

b. crosses the principal axis through the *focal point* refracts along a path that <u>is parallel to this axis</u>.

c. crosses the principal axis through the *center of the lens* <u>is not refracted at all</u>.

d. crosses the principal axis *inside the focal point* refracts along a path that <u>diverges from this axis</u>.

e. crosses the principal axis *beyond the focal point* refracts along a path that <u>converges to this axis</u>.

Materials

☐ Two identical books covered with notebook paper. Fold each sheet along the margin lines, then align each fold with the top of each spine, as before.
☐ A hand lens.
☐ A metric ruler.
☐ Plastic wrap.
☐ A ballpoint pen, straightedge and scissors.

(TO) predict the position of the virtual image of a pin located within one focal length of the lens. To confirm this position by experiment.

VIRTUAL IMAGES ○   Light (   )

1. Crease lined paper down the middle. Draw a *principal axis* along this fold. Outline a side view of your hand lens (about actual size) in the middle, and draw its *vertical axis*.

2. Measure and label the 2 focal points. Draw an observer's eye to the far left.

3. Tape a pin to the principal axis, tilting it as shown. Position its center about one third focal length from the lens.

4. Sketch how rays from the top and bottom of the pin should refract through the lens; then draw where the observer's eye should see them. Justify your drawing.

5. Notch out your lens drawing so a real lens can be inserted, between books. Bob your head up and down between the lens image and its predicted position. Comment on the accuracy of your drawing.

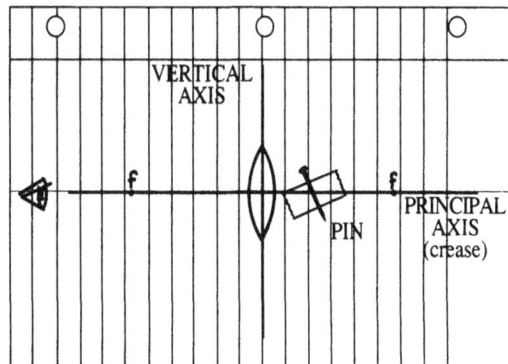

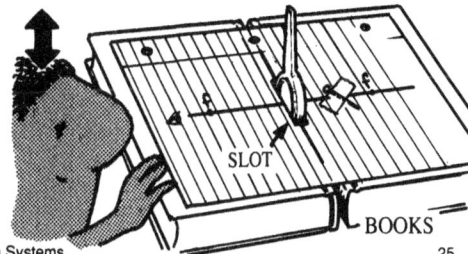

© 1991 by TOPS Learning Systems   25

Introduction

a. Draw the lens diagram below on your blackboard, but exclude the ray lines and rule numbers.

b. Now draw the solid part of both rays, asking students to cite the appropriate rule from the previous activity.

c. Extend the dashed lines back, showing how the eye interprets diverging rays as originating from a common point. Emphasize that this pin head image is *virtual* because light rays don't really come from this point.

d. Ask student volunteers to locate the virtual image of the pin point in a similar manner.

e. Discuss your findings: A pin held halfway between the focus and lens appears enlarged (magnified) and right-side-up (erect).

f. Examine a pin through a hand lens to confirm that this is true.

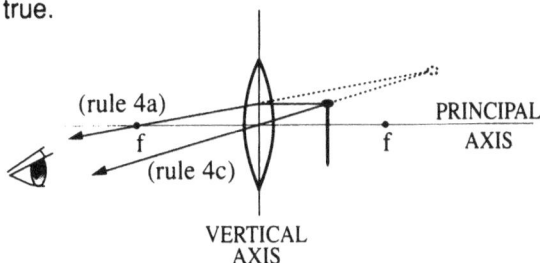

Answers / Notes

4.

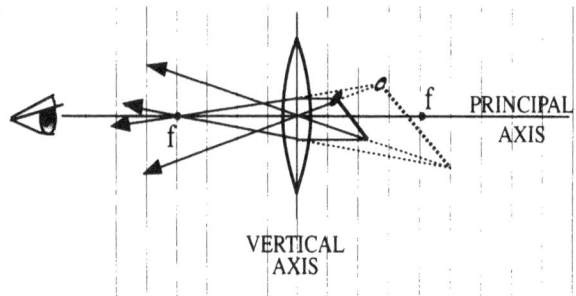

Parallel rays from the pin head and pin point refract through the focal point. Other rays through the center of the lens don't refract at all. The eye traces both sets of diverging light rays back to 2 common points, and there sees the virtual image of the pin head and pin point. All points in between are defined in the same way.

5. If the drawing is accurate, the virtual image of the pin and its plotted position will appear to coincide. Moving the head up and down between these two will cause very little apparent movement.

Materials

☐ Notebook paper.   ☐ A hand lens.
☐ A metric ruler.   ☐ Scissors.
☐ A straight pin.   ☐ Two books of equal thickness.
☐ Clear tape.

(TO) predict the position of the real image of a pin located outside one focal length of the lens. To confirm this position by experiment.

REAL IMAGES　　　　　　　　　　O　　　　　　　　**Light (　)**

1. Cut notebook paper in half lengthwise. Tape the sections end to end.

2. Draw your lens, and a long principal axis down the middle. Measure out 3 focal lengths to each side. Draw in the vertical axis.

3. Tape the points of 2 pins so they touch the principal axis where shown.

VERTICAL AXIS

PRINCIPAL AXIS

3f　　2f　　1f　　　　1f　　2f　　3f

4. Locate the image of each pin head by drawing rays. Draw the rest of each pin to its point on the principal axis.
    a. Describe each pin image.
    b. Tape a pin to the front of your flashlight. Confirm that each image shows up at the predicted distance, by projecting its image through your hand lens onto an index card.

PIN

INDEX CARD

26

Introduction

a. Sketch this diagram on your blackboard. Draw only the lens and axis, plus object pins A and B to start.

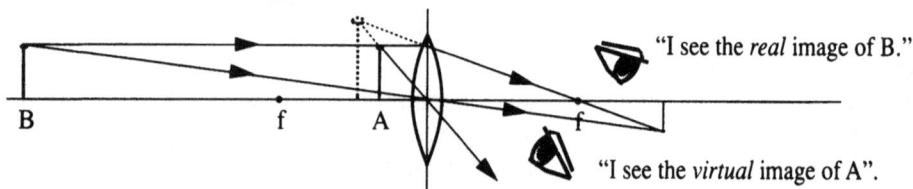

B　　　　　　f　　A　　　　f

"I see the *real* image of B."

"I see the *virtual* image of A".

b. Ask a volunteer to locate the image of pin head A by drawing just 2 rays. *(The eye sees diverging rays from the pin head as originating from a point behind the lens. The pin point extends to the principal axis directly underneath. The image, in dashed lines, is virtual, enlarged and erect.)*

c. Now challenge students to locate the image of pin head B, outside the focal point. *(The rays converge to form a real image that is reduced and inverted. The lines are solid because the image is really there. It can be projected on a screen.)*

Answers / Notes

3. *Placing the pin a little inside 2f allows the real image to enlarge, but still remain within the confines of the paper.*
4.

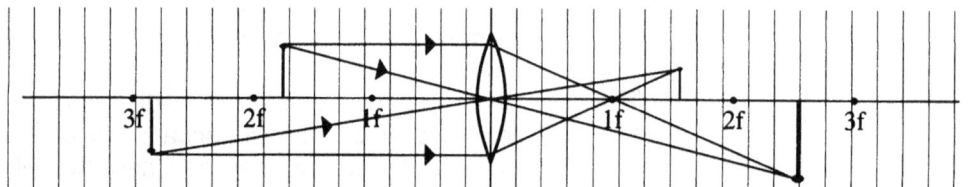

3f　　2f　　1f　　　　1f　　2f　　3f

4a. The image of the object pin inside 2f is real, enlarged and inverted. The pin inside 3f has an image that is real, reduced and inverted.
4b. When the pin, lens and index card are held as diagrammed, each pin image comes into sharp focus.

Materials

☐ Notebook paper, scissors and clear tape.
☐ A metric ruler.
☐ Three straight pins. Avoid extra long pins.
☐ A flashlight, hand lens and index card.

(TO) observe how images through a lens change with an object's distance from the lens. To understand these changes in terms of ray diagrams.

FLIP-FLOP ○ Light ( )

1. Stick a strip of masking tape to the side of a cereal box. Calibrate this tape in focal lengths (of your hand lens), starting at the bottom of the box.

2. Cut out the "object arrows" rectangle. Fold it along the dashed lines and tape it to the bottom of the box, so the arrows extend beside the calibrated tape.

3. Move the hand lens through these focal length positions while looking through it, *at arm's length*, with one eye. Fully describe each arrow image you see, and draw a ray diagram.

CALIBRATED TAPE

OBJECT ARROWS

CEREAL BOX

    a. from 0 f.l. to 1 f.l.
    b. at 1 f.l.
    c. from 1 f.l. to 2 f.l.
    d. at 2 f.l.
    e. beyond 2 f.l.

HOLD THE BOX AT *FULL* ARM'S LENGTH

4. Estimate the power of magnification of your lens: how many times larger can it make the arrows? Explain how you know.

5. Real images float right in space! Capture these arrows on waxed paper. Where should you hold it to see an actual-sized image?

27

Answers / Notes

3. *The box must be held at full arm's length. If not, the eye will accommodate, focusing the arrows on your retina at a closer distance than they would otherwise focus on a screen. This makes the focal length appear longer than it really is.*

We have arranged our ray diagrams with stationary object arrows and variable lens positions to suggest actual experimental conditions. You might outline this basic structure on your blackboard if you wish students to follow a similar format. Otherwise they can draw more traditional ray diagrams, with a fixed lens and moveable object.

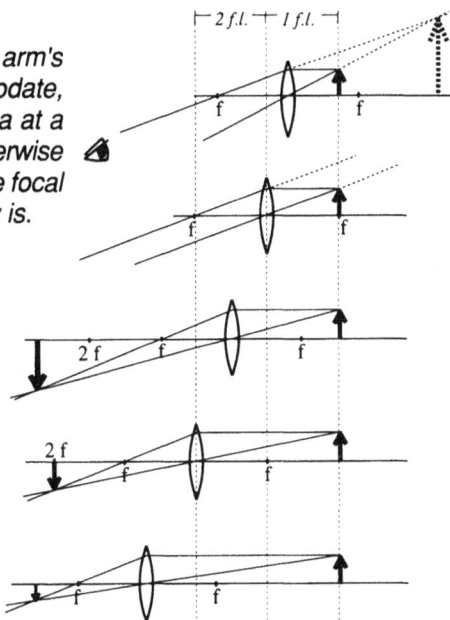

a. The image is virtual, erect and magnified. It continues to grow as the lens approaches 1 f.l.

b. No image forms. The rays neither diverge to a virtual image nor converge to a real one; they move parallel.

c. The image is real, inverted and magnified. It shrinks to near actual size as the lens approaches 2 f.l.

d. The image is real, inverted and actual size, neither magnified nor reduced. *(The image will appear somewhat larger than actual size unless the eye is held* far *beyond 2 f.l.)*

e. The image is real, inverted and reduced. It continues to shrink as the lens moves farther away.

4. Within 1 focal length, most lenses magnify about 3 times with negligible distortion: one arrow viewed through the lens appears to span roughly 3 unrefracted background spaces. *(Students may report much higher magnifications as they move the lens very close to the focal point, but the images distort badly as they increase in size.)*

5. Hold the waxed paper 2 f.l. in front of the lens with the arrows 2 f.l. behind it. *(This is an amazing sight!)*

Materials

☐ Masking tape, scissors and metric ruler.
☐ An empty cereal box.

☐ A rectangle of arrows and lines photocopied from the supplementary page at the back of this book.
☐ A hand lens with predetermined focal length.

(TO) construct working models of a telescope and microscope from two hand lenses.

LENSES IN COMBINATION (1) ○   Light (   )

1. Rubber band waxed paper over your flashlight, and tape a paper clip on the front surface. Mount this on 2 cans with more rubber bands.

2. Bend a paper clip at a right angle. Tape the narrow end to the handle of your lens. Attach a straw cut to the focal length of your lens. Mount it on a battery like this:

3. Mount waxed paper on another battery. Flatten the top to make a screen, as shown.

FLATTEN

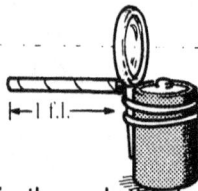

4. Project an image of the paper clip through the lens onto the screen. Magnify the *back* of this screen image with another hand lens.
   a. Describe the first image on the screen; the second that you magnify.
   b. Can you remove the screen and still see this second image? Explain.
5. How should you arrange your equipment...
   a. ...to model a microscope (make the paper clip very large).
   b. ...to model a telescope (see the paper clip at a distance).
      (Save your lens with straw attached for the next activity.)

© 1991 by TOPS Learning Systems   28

Answers / Notes

1. *During this experiment students will be focusing hand lenses on the flashlight head and looking directly into the light. Near the focal point, the concentrations of light entering the eye from an uncovered flashlight are intense and uncomfortable. To diffuse and soften the light, the flashlight head is wrapped in waxed paper. It should remain attached through this entire experiment.*

4a. The first image is real and inverted. *(It may be larger or smaller, depending on which side of 2 f.l. the flashlight head is positioned.)* When viewed through the second lens, the image of the image is magnified and virtual.

4b. Yes. The second lens picks up the real image of the first lens much more clearly without the screen inserted between.

5a. Position the paper clip on the flashlight a little beyond 1 f.l. This projects an enlarged, real, inverted image on the other side of the lens beyond 2 f.l. that can be further magnified by the second lens. *(As the image size increases, the field of view through the microscope becomes smaller, and possibly lost, unless properly aligned. When aligned, the magnification is impressive.)*

5b. Position the paper clip on the flashlight at a distance, far beyond 2 f.l. This projects a reduced, real, inverted image on the other side of the lens just beyond 1 f.l. that can then be magnified by the second lens. *(Because both lenses have the same focal length, the final virtual image is the same size as the original object. Students will improvise a telescope beyond 1 power in the next activity.)*

Materials

☐ Rubber bands.
☐ Waxed paper and scissors.
☐ A flashlight.
☐ Clear tape.
☐ Paper clips.

☐ Two medium-sized cans, about 15 ounces.
☐ Two hand lenses.
☐ Masking tape.
☐ A straw.
☐ Two size-D batteries.

(TO) construct and use more working models of telescopes and microscopes.

LENSES IN COMBINATIONS (2) ○   Light (   )

1. Tape the narrow ends of bent paper clips to 4 hand lenses like these. (Note: 2 lenses are taped together.)

SINGLE LENS
(1 paper clip)

SINGLE LENS
(2 paper clips)

DOUBLE LENS
(1 paper clip)

2. Cut straws to these focal lengths.

— 1 f.l. —

— 2 f.l. —

EXPERIMENT!

3. Connect these combinations. Report your findings:
   a. **Inverter:** single lens + long straw + single lens.
   b. **Telescope:** double lens + short straw + single lens.
   c. **Microscope:** double lens + short straw + single lens + long straw + single lens.

29

Answers / Notes

1. *The first lens was already prepared in step 2 of the previous activity. As before, paper clips should all be bent to right angles. Then the narrow end of each clip should be taped to each handle just below the lens.*

2. *In step 2 of the previous activity students have already cut a straw to 1 focal length (8.2 cm for hand lenses sold by TOPS). Here a second segment should be cut twice as long.*

3a. The single lenses should be separated 2 focal lengths (the length of the longer straw). Looking through either lens, distant images are inverted, but neither enlarged nor reduced.

3b. The double and single lenses should be separated by *more* than 1 focal length! Do this by sliding the paper clips at each end part way out of the straw. Using the double lens as the eyepiece and the single lens as the objective, distant images are inverted and somewhat enlarged. (Looking at a wall clock across the room, for example, with both eyes open, students will see a virtual image about twice as large as the real clock.) Looking through this telescope in reverse inverts and reduces distant images to about half size.

3c. The double and single lenses need only be separated by 1 focal length. (A little longer is OK too.) And both single lenses should be separated by 2 focal lengths. Holding this assembly about 1 focal length from any object, while looking through the double lens, gives strong magnification and a wide field of view.

Materials

☐ Four hand lenses of equal focal length. One is already fitted with a 1 f.l. straw.
☐ Four paper clips.
☐ Masking tape.

☐ Scissors.
☐ Straws.
☐ A metric ruler.

(TO) observe how increasing lens curvature shortens the focal length. To use this principle to explain how your eye focuses images on its retina.

WHERE IS THE FOCUS? (1)   ○   Light (   )

1. Rubber band plastic wrap over a canning ring. Pull on the edges to make it wrinkle free, then deposit a line of different-sized water drops *inside* with a paper clip.

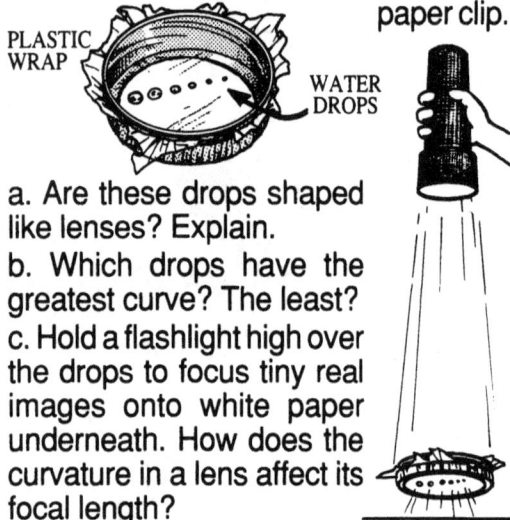

PLASTIC WRAP

WATER DROPS

a. Are these drops shaped like lenses? Explain.

b. Which drops have the greatest curve? The least?

c. Hold a flashlight high over the drops to focus tiny real images onto white paper underneath. How does the curvature in a lens affect its focal length?

2. To clearly see this point at arm's length, the eye must focus its diverging rays to a corresponding point on your retina.

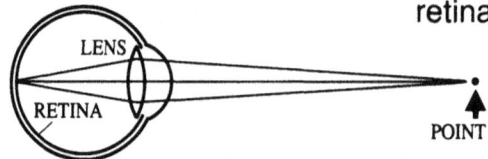

LENS

RETINA

POINT

a. Can you see this same point clearly only a hand-span away? Redraw the diagram, to explain how the eye accommodates (refocuses) to this shorter distance.

POINT

b. Can you see this point clearly only a thumb-width away? Redraw the eye diagram to illustrate.

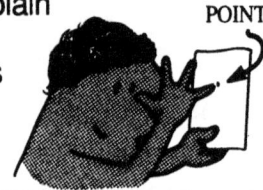

© 1991 by TOPS Learning Systems   30

Answers / Notes

1a. Yes. They are flat underneath and curved on top *(plano-convex)*.

1b. As drops decrease in size, their surfaces curve more tightly.

1c. As lens curvature increases, light rays are refracted through sharper angles. The focal length decreases.

2a. Yes. The eye accommodates to this shorter distance by increasing the curvature in its lens. This refracts the light rays (increasingly divergent as the point approaches) through sharper angles, maintaining the focus at the retina.

2b. No. The eye muscles cannot increase the curvature of the lens sufficiently to focus extremely divergent rays of light from this point to a corresponding point directly on the retina.

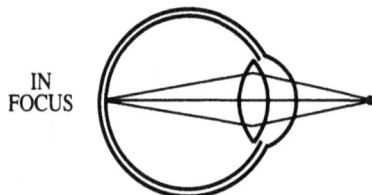

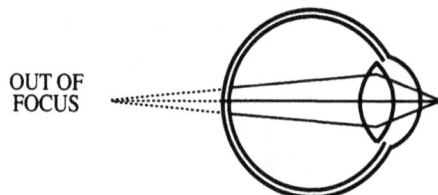

IN FOCUS

OUT OF FOCUS

Materials

☐ A rubber band.
☐ Plastic wrap.
☐ A canning ring. The circle of metal that holds a canning lid tightly to its jar. Purchase this in the home canning section of your grocery store.
☐ A jar of water.
☐ A paper clip.
☐ A flashlight.

(TO) simulate near-sighted vision with a hand lens, and refocus the blurred image with a pinhole. To understand why pinholes keep light in focus at all distances.

WHERE IS THE FOCUS? (2) ○   Light (   )

1. Rubber band waxed paper to the front of your flashlight. Fix a pin to it with clear tape.

2. If you wear glasses, remove them. Hold a hand lens directly against your eye while looking at the lighted pin held at arm's length.
   a. How does it look?
   b. Draw a ray diagram that shows why this lens makes you near-sighted.

HAND LENS CLOSE TO EYE
PIN ON WAXED PAPER

3. Poke a *very* tiny hole with the tip of a pin in a small piece of foil. Press this hole between the hand lens and your eye. Now look at the lighted pin at arm's length, as before.
   a. How does the pin look?
   b. Redraw your ray diagram from step 3, to show why the pinhole refocuses the fuzzy lens image.

FOIL WITH TINY HOLE

4. Near-sighted people sometimes squint through nearly closed eyes to improve their distance vision. Try this while pressing the hand lens to your eye. Explain why this helps.   (Keep your flashlight covered for the next activity.)

© 1991 by TOPS Learning Systems   31

Answers / Notes

2. *By holding the hand lens directly against (but of course, not touching) the eye, a normal-sighted person can simulate extreme myopia. Students who are already nearsighted will see an unfocused image whether or not they use the hand lens.*

2a. The pin image is extremely blurred. At arm's length it is hardly visible at all.

2b. Light from the pin refracts inward through both the hand lens and eye lens, focusing in front of the retina instead of on its surface.

3a. The pin is now in much better focus. At arms length, it is quite clear. Less light is admitted, however, so it looks darker.

3b. The pinhole admits only a single narrow beam of light. Each point on the pin strikes the retina at a corresponding point, thus providing a relatively clear image. All other rays, that would otherwise blur the pin image, are blocked by the foil.

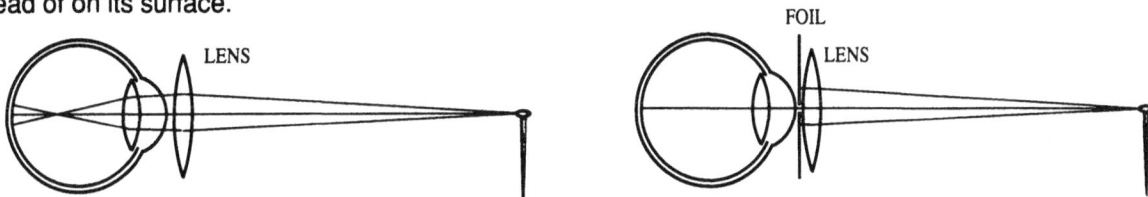

LENS

FOIL / LENS

4. Blurred distant images will clear somewhat when the eye is almost closed in a hard squint. In effect, your closed lids and lashes block, much like a pinhole, most of the light rays that would otherwise focus too far in front of the retina, leaving a narrower angle of beams to stimulate the retina in a somewhat focused image.

Materials

☐ A flashlight.
☐ Waxed paper.
☐ A rubber band.
☐ Clear tape.
☐ Straight pins.
☐ A hand lens.

(TO) explore three different ways to magnify the image of a pin. To compare and contrast each method.

THREE MAGNIFIED VIEWS ○   Light ( )

1. Slowly bring the pin (taped to your lighted flashlight) in from arm's length closer to your eye.

a. Why does it look larger and larger?
b. Why does the image eventually blur?
c. Could you see the pin more clearly at close range through a pinhole? Make a prediction.
d. Test your prediction with different sized holes.

TINY HOLE: MEDIUM HOLE: FULL PINHOLE:

PAPER PADDING   FOIL

2. Rubber band plastic wrap over a canning ring, as before, so it is wrinkle free. Deposit drops of various sizes on the *outside* with a paper clip, then look through them at your lighted pin.

a. Which drops magnify the most?
b. Are the images in steps 1 and 2 real or virtual? Explain.

3. Remove the waxed paper and tape the pin directly to the flashlight head. Hold the hand lens *just beyond* 1 focal length, projecting the pin image onto an unlit wall.

WALL

a. How can you change the magnification?
b. Compare images in steps 2 and 3.

© 1991 by TOPS Learning Systems   32

Answers / Notes

1. *If the light is too bright for comfort at close range, students may add an additional layer or two of waxed paper.*

1a. The pin looks larger as you bring it closer because it continually fills greater portions of your field of vision.

1b. The pin blurs when the eye lens can no longer refract the light sufficiently to focus it on the retina.

1c. Yes. Since the pinhole allows only a single beam of light to enter the eye from each point on the pin, the image should remain in focus at all distances.

1d. The pinhole permits the eye to register a focused retinal image of the pin at extremely short distances, thus filling nearly your entire field of vision. The smaller this hole, the clearer (but dimmer) this image.

2. *Very tiny drops of water produce a remarkably high power of magnification. However, the eye must be positioned at very close range directly over the drop. Because accidental contact between the eye and the water drop is possible, be sure to use clean water.*

2a. The smallest drops (with greatest curvature and shortest focal length) give the greatest magnification.

2b. The image in step 1 is real. Light enters the eye from where the object pin is actually located. It looks large because it is very close. The image in step 2 is virtual. Light is refracted through the water drop so it appears to originate from a much larger virtual pin behind the object pin.

3. *It is possible to project huge images across the length of your room if it is sufficiently dark. Try substituting a dead insect, or just its wing, for the pin. The view is impressive.*

3a. Change the magnification by changing the distance between the pin and lens. Approaching 1 focal length, the pin image grows enormously.

3b. In step 2, the pin image is virtual and erect. In step 3, it is real and inverted.

Materials

☐ The flashlight with pin and waxed paper from the previous experiment.
☐ Aluminum foil.

☐ A pin.
☐ A rubber band.
☐ Plastic wrap.
☐ A canning ring.

☐ A jar of clean water. See note 2 above.
☐ A paper clip.
☐ A hand lens.

☐ A light-colored wall without direct illumination. If necessary, turn off most lights or pull down the window shades.

(TO) reproduce a reflected mirror image on paper, using the principle that glass both reflects and transmits light. To study the reversed nature of the reflected virtual image.

REVERSE IMAGES   O   Light (   )

1. Write your name or initials in large capital letters between two lines of notebook paper.

2. Hold a microscope slide *upright* on the top line. Look *down* through the slide at a steep angle, tracing the reflection on the other side.

3. Move the slide up one line at a time, drawing reflections of reflections, until you complete four lines. What pattern have you created?

4. Poke your pencil through a circle of index card so you can see the point only in a mirror.

    a. Write your name so its mirror image appears normal.
    b. Why is this so hard to do?

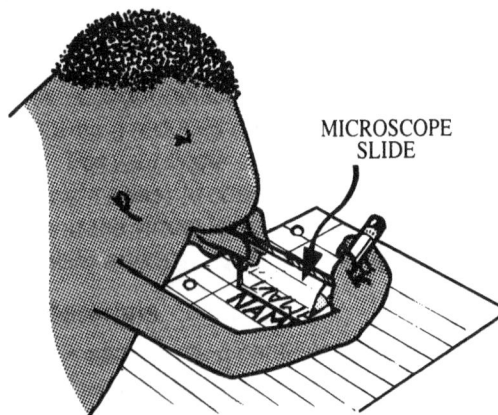

MICROSCOPE SLIDE

INDEX CARD SHIELD

MIRROR

33

Answers / Notes

3a. Each reflected image is a reversal of the previous reflection: the letters start right-side-up and frontwards, then change to upside-down and backwards. This pattern then repeats.

NAME
NAME
NAME
NAME

4a.

4b. Looking into a mirror, the brain must coordinate hand movements with a reversed visual image. To move the *image* of your pencil *down* the paper, for example, you must actually move your *pencil up* the paper. Confusion results because your hand must do the opposite of what your eye sees happening.

Materials

☐ Notebook paper.
☐ A microscope slide.
☐ A rectangular plane mirror.
☐ An index card.
☐ Scissors.

(TO) study symmetry in letters of the alphabet using a plane mirror.

SYMMETRY　　　　　　　　　　○　　　　　　　　**Light (　)**

1. Use a mirror to find which letters are symmetrical — can be divided into similar halves. Summarize your findings in the table below, using each letter just once.

"A" has symmetry about its vertical axis, but not about its horizontal axis.

A B C D E F G H I J K L M N O P Q R S T U V W X Y Z

no symmetry	
vertical axis of symmetry only	
horizontal axis of symmetry only	
vertical and horizontal axes of symmetry	

2. Draw a letter "X" that has *four* axes of symmetry. How did you do this?
3. Alter the letter "O" to make over a million axes of symmetry! Explain.
4. Spell a word that has horizontal line symmetry; vertical line symmetry.

　　34

Introduction

Cut out a large capital letter "A" from a piece of paper.

a. Demonstrate its *symmetry* along the vertical axis: fold it in half or place a mirror on the vertical axis to show that both sides are similar.

b. Demonstrate its *asymmetry* along the horizontal axis: fold it in half horizontally or use a mirror to show that the upper and lower parts are different.

Answers / Notes

1.

no symmetry	F G J K L N P Q R S Z
vertical axis of symmetry only	A M T U V W Y
horizontal axis of symmetry only	B C D E
vertical and horizontal axes of symmetry	H I O X

2. Draw an "X" with lines that meet on center at right angles. This X has 2 additional axes of symmetry: one along each diagonal.

3. If the "O" is drawn in the shape of a perfect circle, it has symmetry along all of its infinite number of diameters. *(This is known as radial symmetry.)*

4. Horizontal symmetry: HIDE, CHOICE　　　　Vertical symmetry: TOOT, MOM

Extension

Print the phrase "FIRST CHOICE" in large, neat capital letters on a piece of paper, writing "FIRST" in pencil and "CHOICE" in pen. Cap a test tube full of water with a cork or lump of clay. Look through this "lens" at both words in the phrase. Ask why "FIRST" appears inverted but not "CHOICE". *(No, it's not because the words are written in pencil and pen!)*

Materials

☐ A plane mirror.

(TO) study how images are projected by concave and convex mirrors.

CURVED MIRRORS (1)   O   Light (   )

1. First draw a *semicircle* with a *principal axis*, and label the *center* of curvature C. This represents a concave mirror.

2. Next stand an object *arrow* on the axis beyond C, and extend a *parallel ray* from its tip to a *point* on the mirror.

3. Draw a *normal* through C to this point, and a *tangent* perpendicular to the normal.

4. Think of this tangent as a plane mirror. Draw the *reflected* ray using a protractor. Label the *focal length* where it intersects the principal axis.

5. Draw another ray from the tip of the object arrow through C. Because this ray reflects back on itself, you can locate the tip of the image arrow. Describe it.

6. Repeat steps 1-5 for a convex mirror of similar size. Describe the resulting image.

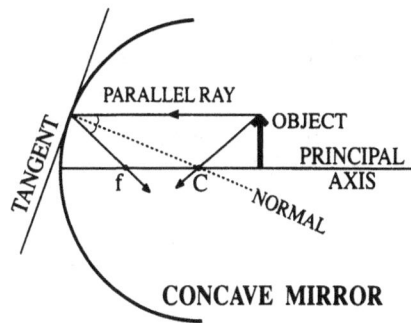

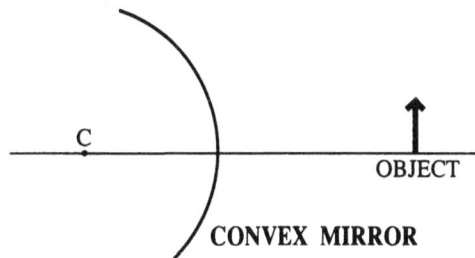

35

Answers / Notes

1. *Encourage students to draw a much larger diagram than shown.*
4. *Students should recall here that ∠i = ∠r.*
5. The image is reduced, inverted and real.

6. *In order to locate the image of the arrow's tip in steps 4 and 5, students should extend the reflected diverging rays to a point behind the mirror.*
   The image is reduced, erect and virtual.

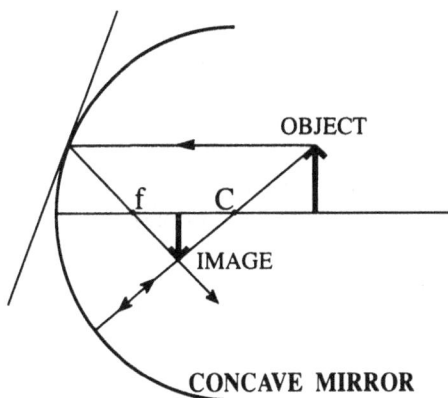

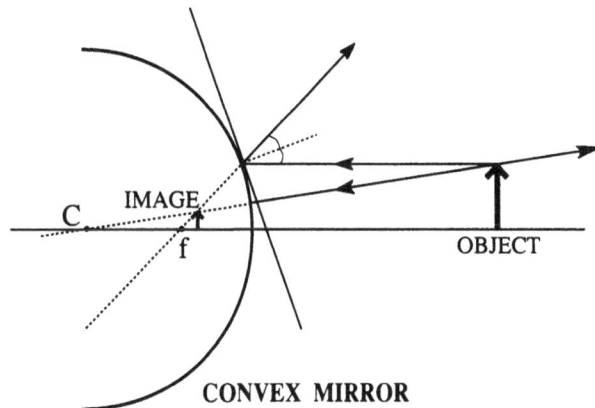

Materials

☐ A protractor. The one previously cut from paper is suitable.
☐ An index card or other straightedge.

(TO) confirm that concave and convex mirrors produce images as predicted in the previous activity.

CURVED MIRRORS (2)　　　　O　　　　　　　　Light (　)

1. Wrap a spoon in a *single* layer of foil, shiny side up. Trim, leaving a small margin to anchor around the edge.

2. Vigorously polish the foil, front and back, with tissue, so you can easily see the reflection of your head and shoulders.

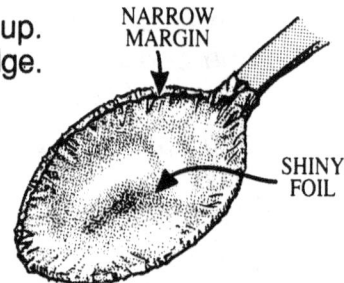

NARROW MARGIN

SHINY FOIL

   a. Is your image on the concave side as your previous drawing predicts? Explain.
   b. What about the convex side?

3. Cup the concave side of the spoon over your eye. This places your eye *inside* the spoon's focal length. Let in enough light from the side to view the image.

   a. Describe what   b. Draw a ray diagram to show how this image forms.
     you see.

4. Hold a shining flashlight in front of a hand lens like this:

   a. Are the two "floating" images you see refractions or reflections? Why?
   b. Which is real and which is virtual? Use waxed paper to find out.
   c. How are these images formed?

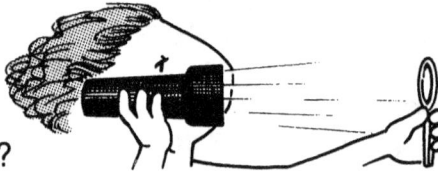

36

Answers / Notes

2a. The image on the concave side is reduced and inverted, as predicted.

2b. The image on the convex side is reduced and erect, as predicted.

3a. An enlarged, erect eye reflects from the spoon at this close range.

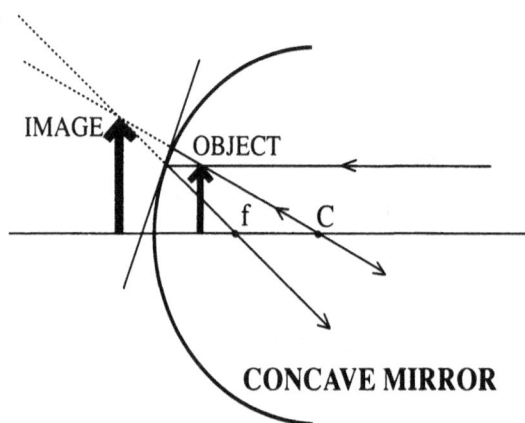

3b.

CONCAVE MIRROR

4a. Both images must be reflections. If not, light would pass through instead of rebounding into the eye.

4b. The front image is real because it can be projected onto waxed paper. The back image is virtual because it can't be projected. *(Put your hand on the back of the lens and this back image virtually appears to shine from under your skin.)*

4c. Because the lens reflects at least some of the light that strikes it, it functions, in this respect, like two curved mirrors. Light reflecting from the convex *front* surface forms a reduced virtual image behind the lens. Light reflecting from the concave *back* surface forms a reduced real image in front of the lens.

   Put your finger on the face of the shining flashlight and it's easy to observe that the real image is inverted and the virtual image is erect. All of these properties were anticipated in the drawings of activity 35.

Materials

☐A spoon with a continuously curving head. It need not be spherical, but must not have a flat bottom.
☐Aluminum foil.
☐Scissors.
☐A facial tissue or piece of soft toilet tissue.
☐A hand lens.
☐A flashlight.
☐Waxed paper.

REPRODUCIBLE
STUDENT
TASK CARDS

Task Cards Options

Here are 3 management options to consider before you photocopy:

1. Consumable Worksheets: Copy 1 complete set of task card pages. Cut out each card and fix it to a separate sheet of boldly lined paper. Duplicate a class set of each worksheet master you have made, 1 per student. Direct students to follow the task card instructions at the top of each page, then respond to questions in the lined space underneath.

2. Nonconsumable Reference Booklets: Copy and collate the 2-up task card pages in sequence. Make perhaps half as many sets as the students who will use them. Staple each set in the upper left corner, both front and back to prevent the outside pages from working loose. Tell students that these task card booklets are for reference only. They should use them as they would any textbook, responding to questions on their own papers, returning them unmarked and in good shape at the end of the module.

3. Nonconsumable Task Cards: Copy several sets of task card pages. Laminate them, if you wish, for extra durability, then cut out each card to display in your room. You might pin cards to bulletin boards; or punch out the holes and hang them from wall hooks (you can fashion hooks from paper clips and tape these to the wall); or fix cards to cereal boxes with paper fasteners, 4 to a box; or keep cards on designated reference tables. The important thing is to provide enough task card reference points about your classroom to avoid a jam of too many students at any one location. Two or 3 task card sets should accommodate everyone, since different students will use different cards at different times.

LIGHT AS PARTICLES ⭘ Light (   )

1. Add a *tiny* pinch of powdered milk to a small jar of water. Shake it vigorously. It should look only slightly cloudy.

SLIGHTLY CLOUDY SUSPENSION

2. Wrap the jar in a dark cloth while shining a flashlight through a small opening. Look through the top and record your observations.

DARK CLOTH

3. Tape a piece of foil to the jar, shiny side in. Repeat the experiment, directing the beam through the water at the foil. What do you see?

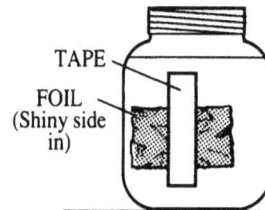

TAPE
FOIL (Shiny side in)

4. It is useful to think of light as composed of incredibly tiny, fast moving particles (*photons*). How do these photons appear to move?

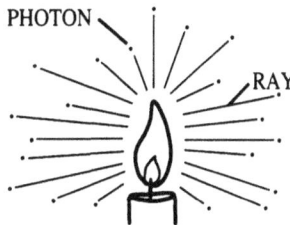

PHOTON
RAY

5. The path a light photon takes is called a *ray*. Draw a light ray diagram to illustrate why you can't see your ear without a mirror.

© 1991 by TOPS Learning Systems   1

PINHOLE VIEWER (1) ⭘ Light (   )

1. Get 2 tin cans of equal size with both ends removed. Cover just one with foil at one end, and rubber band waxed paper to the other end.

RUBBER BAND
SECOND CAN UNCOVERED
FOIL
WAXED PAPER

2. Poke a pinhole in the center of the foil. Enlarge it to the size of a pin head with a sharp pencil point.

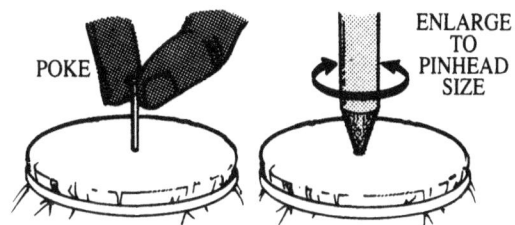

POKE
ENLARGE TO PINHEAD SIZE

3. Hold the uncovered can over the waxed paper end, while looking through it toward a well-lit area.

    a. Write your observations.
    b. Is the pinhole image erect (right-side-up) or inverted (upside-down)?
    c. Draw a labeled light ray diagram to explain your observations.

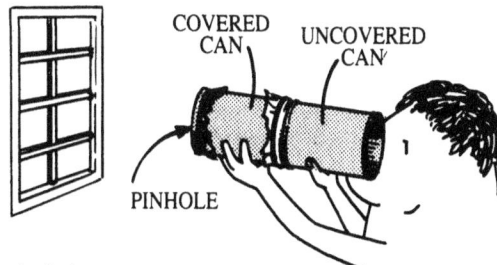

Save your pinhole viewer to use again.

COVERED CAN
UNCOVERED CAN
PINHOLE

© 1991 by TOPS Learning Systems   2

PINHOLE VIEWER (2) ◯ **Light (   )**

1. Look at a shining flashlight through your pinhole viewer. Use ray diagrams to explain each observation.

   a. As you move the object close or far away, the pinhole image grows or shrinks.

   b. As you move the object in one direction, the pinhole image moves in the opposite direction.

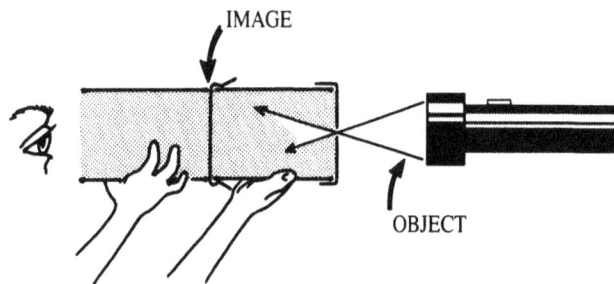

IMAGE

OBJECT

2. Think of a way to put multiple images of the same object on your screen. Explain how you did this.

3

SHADOW DISK ◯ **Light (   )**

1. Tape white paper to an empty cereal box, forming a screen.

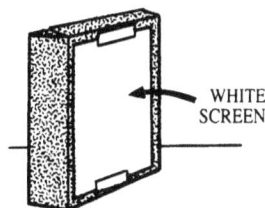

WHITE SCREEN

2. Cover the end of your flashlight with foil. Poke a hole in the middle, about as big as your pencil.

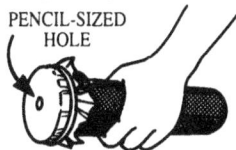

PENCIL-SIZED HOLE

3. Trace a circle around a battery onto an index card. Cut it out and attach a paper clip handle.

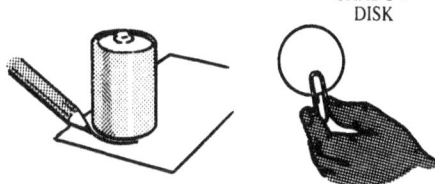

SHADOW DISK

4. Project shadows of this disk on your screen with the flashlight. Draw ray diagrams explaining how to make the shadow larger or smaller.

5. A distant light source, like the sun, produces light rays that strike the earth nearly parallel.

   a. Will the size of your disk's shadow change if you hold both it and the screen perpendicular to the sun's rays? Make a prediction.

   b. Test your prediction when the sun is shining.

RAYS FROM THE SUN

Save your screen and shadow disk to use again.

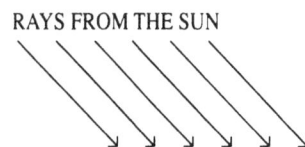

4

THE SHADOW IS FUZZY!   O   Light (   )

1. Mount a flashlight on 2 cans with rubber bands. Replace the foiled end with waxed paper.

2. Mount your shadow disk on a battery with a rubber band. Align with your screen in a dim place to project a shadow.

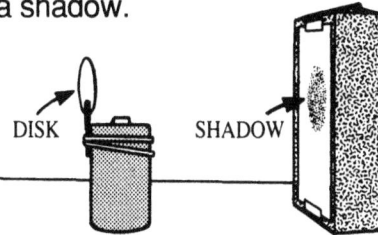

DISK   SHADOW

WAXED PAPER

3. Shadows are usually dark in the middle (the umbra) and lighter around the edges (the penumbra). Explain how to make a shadow that is...

   a. nearly all umbra.
   b. nearly all penumbra.
   c. an equal portion of umbra and penumbra.

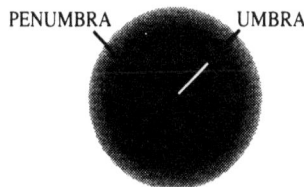

PENUMBRA   UMBRA

Save your equipment for the next activity.

4. Draw light rays from points x and y to explain why...

x
y
FLASHLIGHT   DISK   SCREEN

   a. an umbra and penumbra both form on the screen.
   b. the percentage of umbra increases as you move the disk nearer the screen.

5

SOLAR ECLIPSE   O   Light (   )

1. Punch a row of five holes across an index card, as far in as the punch will reach. Attach to a battery with a rubber band and paper clip.

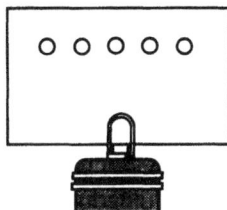

2. Project a small umbra surrounded by a large penumbra on your screen. Position the index card so its holes span the complete shadow.

3. Now remove the screen to look at your "eclipse of the sun" through each hole. (Never do this with the real sun. It will damage your eyes.)

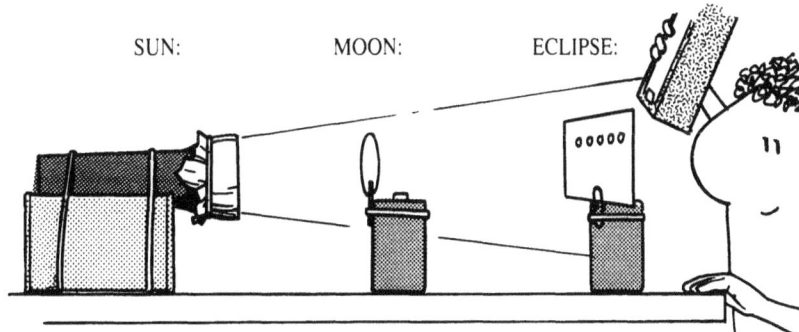

o o o o o

SUN: MOON: ECLIPSE:

4. Draw the parts of the "sun" and "moon" you see through each hole.
   a. Relate each part of this model to a real eclipse.
   b. Which drawing(s) show(s) that you are standing in the penumbra? In the umbra? Explain.

6

REFLECTION   O   Light (   )

1. Cut around the box containing the protractor. Fix it to your table with masking tape.

2. Tie a small loop in the middle of some thread, then tape its knot exactly where the normal meets the baseline. Tape the ends to pennies labeled *i* (for *incident ray*) and *r* (for *reflected ray*).

3. Lightly stick the back of a plane mirror to a battery with tape rolled sticky-side-out. Set the mirror directly on the baseline.

4. Slide the incoming ray, i, to any angle with the normal. Line up the outgoing reflected ray, r, with the reflection you see in the mirror.

    a. Compare the angle of incidence ($\angle$ i) with the angle of reflection ($\angle$ r).
    b. Does this relationship hold for all angles? Try and see.
    c. What is the value for $\angle$ i and $\angle$ r when you look at the pupil of your eye?
    d. Is it possible to reflect a ray for $\angle$ i = 90°?

5. Place a paper clip at $\angle$ i = 40°. Predict where you should hold a straw to see the image of this paper clip projected through it. Try it and see.

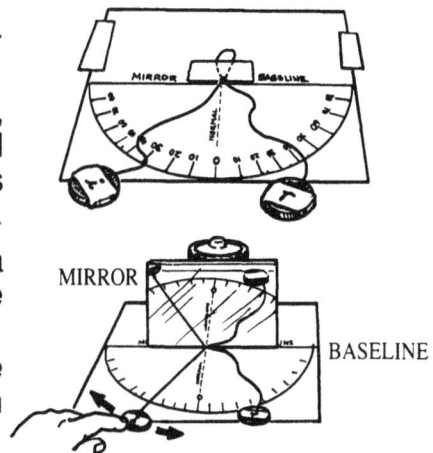

7

LINE UP (1)   O   Light (   )

1. Darken the middle line on notebook paper. Tape the *edge* of a mirror to a battery, then set its reflecting surface vertically over this baseline.

2. Cut a straw in half. Rubber band each half to other batteries so they stand perfectly straight.

3. Set *straw Y* at some distance behind the mirror. Position *straw X* somewhere in front of the mirror so its mirror image lines up perfectly with straw Y behind. The alignment is correct when you can move your head from side to side and still see only one straw behind the mirror.

4. Count the lines from each straw to the baseline. What seems true?

5. Repeat the experiment with straw Y placed closer to the mirror. Do you get a similar result?

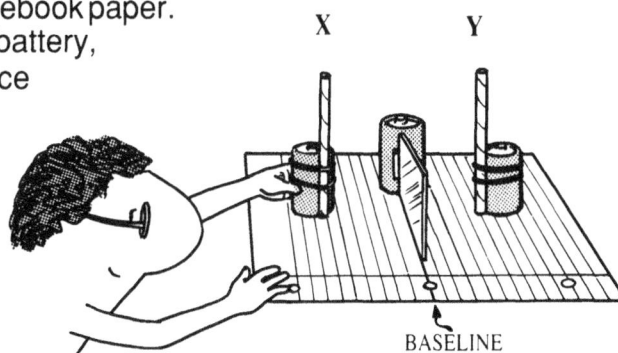

8

LINE UP (2)   〇   Light (   )

1. Fold notebook paper in half lengthwise. Mark the crease to form a baseline. Draw an x below, exactly on a notebook line. This represents an object and a mirror.

2. Draw an incident light ray from this x along the normal. Draw two more rays, each meeting the mirror at intersecting lines.

3. Show how these rays reflect so ∠i = ∠r. Accurately draw and label all angles.

4. Extend the reflected rays behind the mirror using dashed lines. Where they cross locates the mirror image of x.

INCIDENT LIGHT RAYS

BASELINE

MEET AT LINES, NOT SPACES

5. Stand a mirror on the baseline. Is the reflected x positioned where you predicted it should be? Explain.

6. Repeat this experiment with a triangle, drawing the fewest possible rays to locate its image behind the baseline. Again, check your accuracy with a real mirror.

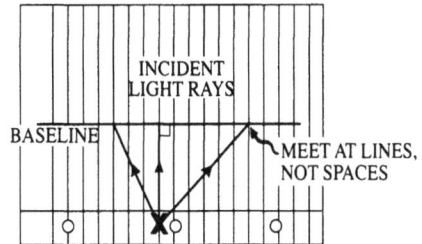

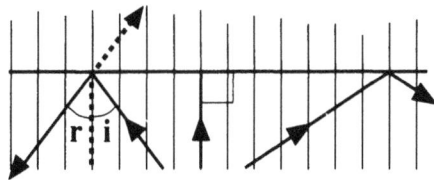

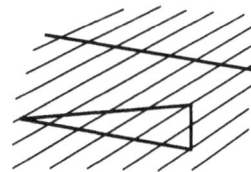

9

UP OR DOWN?   〇   Light (   )

1. The image behind a plane mirror is said to be *virtual.* Light doesn't really come from the candle behind the mirror, but it looks virtually like it does.

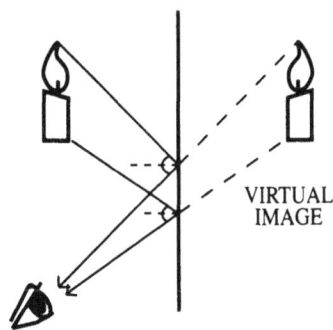

VIRTUAL IMAGE

a. Does the eye interpret light rays as bouncing off the mirror? Why are dashed lines used in this diagram?

b. Set a battery on your head and look at its reflection in a mirror. Draw a ray diagram to show the relative positions of your eye, the battery and its virtual image.

2. Position 2 mirrors to observe what is behind you. Flip the *top* mirror to reflect what is ahead.

LOOKING BEHIND

LOOKING AHEAD

a. How is the scene behind you different from the one ahead?

b. Account for these differences using ray diagrams.

10

FUNNY FACES   ○   Light (   )

1. Slide 2 mirrors together so their middle junction runs through the image of both your eyes.

2. Put a penny under the edge of each mirror to give your face four eyes.

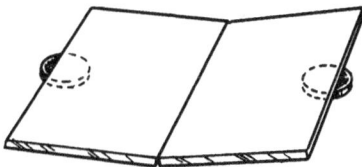

3. Slide the pennies under the crack in the middle of the mirrors to give yourself no eyes!

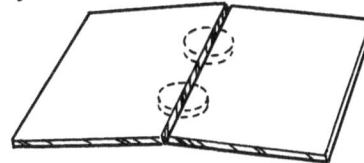

4. Use diagrams to explain the illusions in steps 2 and 3.

11

VIRTUAL REALITY   ○   Light (   )

1. Cover the head of your flashlight with foil. Poke a pencil hole in the center.

2. Attach the long edge of a glass microscope slide to a battery with masking tape.

3. Fill a jar with water. Use all of these objects to make a spot of light appear as if it were virtually floating under water.

4. Tell how you created this illusion. Explain how it works.

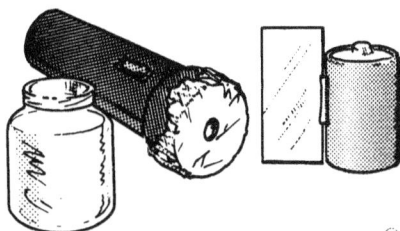

5. Put a penny on white paper. Hold a glass slide on edge just behind it.

6. Tip the glass slide forward so you can see the image of the penny reflected in the slide.

7. How should you move the slide to make a penny image first appear, then fade and disappear.

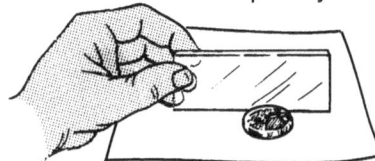

8. Light both reflects off glass and passes through it in amounts that depend on $\angle i$. Use this concept to explain why the penny fades.

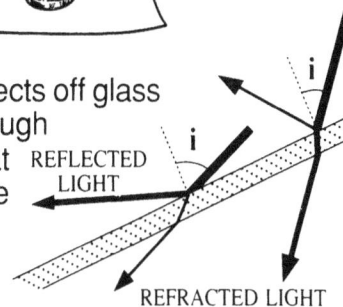

REFLECTED LIGHT

REFRACTED LIGHT

12

LIGHT AS WAVES ○ Light ( )

1. Make a ripple tank: add enough water to a rectangular cake pan to *just* cover a straw.

2. Drip water from an eyedropper high above the center of the calm water.
   a. How do water waves travel outward from the drop?
   b. Do water waves reflect off all sides of the pan?
   c. Imagine striking a match in a totally dark room to see your surroundings. How do water waves in a pan model this event?

3. Trim the straw (if necessary) to easily fit inside the width of the tank. Add a masking tape handle in the middle of the straw.
   a. Make waves by moving the straw up and down at one end of the tank. Do these waves reflect? Does $\angle i = \angle r$?
   b. Now make waves by moving the straw up and down at the *corner* of the tank (diagonally). Do these waves reflect? Does $\angle i = \angle r$?

(Save your taped straw.)

13

REFRACTION (1) ○ Light ( )

1. Set up your ripple tank on a flat surface as before, with enough water to just cover your wave-generating straw.

2. Now tip the pan by sliding a pencil under one edge. Generate *wave fronts* as before and record your observations.

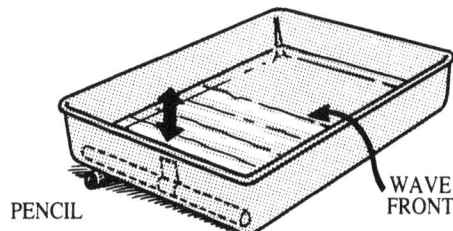

PENCIL
WAVE FRONT

3. Waves slow down in shallow water due to friction against the bottom.
   a. Explain how this bends an advancing wave front.
   b. Bend the wave front in the opposite direction. Explain how you did this.

4. Rubber band 2 batteries to a flashlight. Rest this at an angle on an open can so it shines a spot of light against the bottom of the emptied pan.

5. Mark the near edge of this light spot with a paper clip. How does this spot shift as you add more and more water to the pan?

6. Light waves "drag" more slowly through water than air. How does this fact explain what you see?

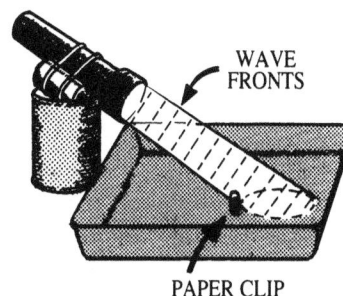

WAVE FRONTS

PAPER CLIP

14

cards 13-14

REFRACTION (2)   O   Light (   )

1. Place 4 glass slides on edge. Draw an outline around them on paper.

2. Draw and label a line to represent an incoming incident light ray as shown.

3. Sight through the glass with an index card where this incident ray emerges on the other side. Draw this line of sight and label it the refracted (bent) light ray.

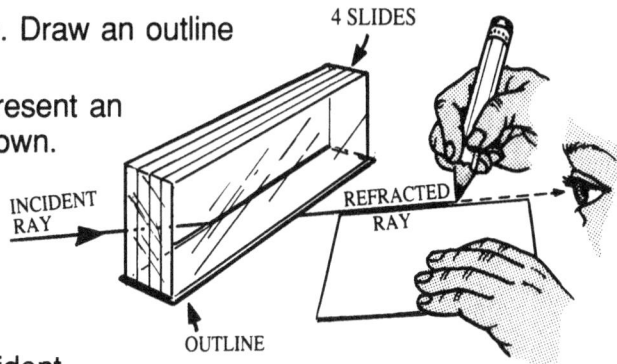

4. Remove the slides. Join the incident and refracted light rays with another straight line. Label the points of refraction.

5. Think of your pencil as an advancing wave front. Does light travel faster or slower as it passes through glass? Explain.

6. Which property of light do you think best explains refraction — its particle nature or wave nature?

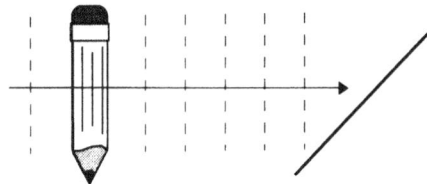

15

REFRACTION (3)   O   Light (   )

1. Think of your pencil as an advancing wave front. Describe how light should refract in each group of 4 slides. Explain your reasoning.

a.   b.   c.

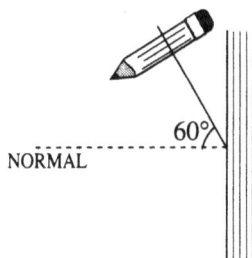   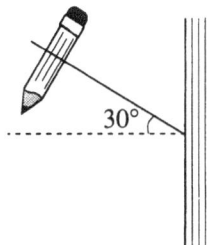   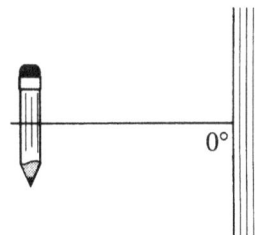

2. Test each experiment by drawing incident light rays at the correct angles and sighting through the slides as before. Were your predictions correct?

3. Pivot the 4 slides, as a group, over lines of notebook paper like this. Do the lines appear to move as you predicted above? Explain.

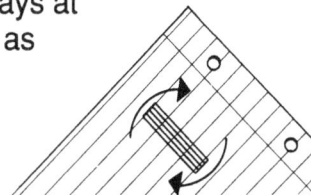

16

WHERE'S THE PENNY?   O   Light (   )

1. Place a penny at the bottom of a shallow can. Position your eye so the penny is just out of sight.

    a. Without changing your line of sight, add water to the can. What do you notice?

    b. Explain your observations with a diagram.

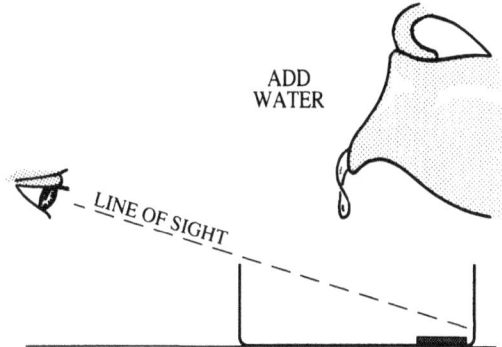

2. With the can full of water, try to "spear" the penny with a straw.

    a. If you aim directly at the penny, can you hit it?

    b. Explain your observations with a diagram.

ADD WATER

LINE OF SIGHT

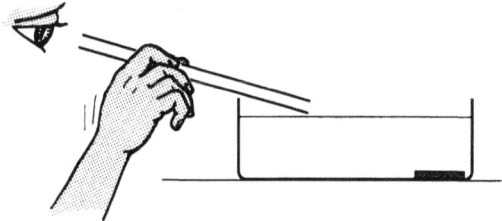

17

WATER LENS   O   Light (   )

1. Wrap a rubber band twice across the mouth of a small jar.

2. Push 2 small test tubes into the jar between the bands.

3. Unbend 2 paper clips in the middle, then pull out the large end just a little so they wedge inside each tube, about 1/3 down from the rim.

4. Fill both tubes with water; the jar half full.

5. Use an eyedropper to transfer water between the tubes and jar. Tell how to make these different surface shapes in the tubes:

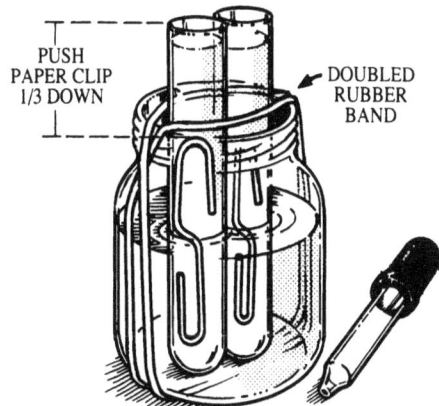

PUSH PAPER CLIP 1/3 DOWN

DOUBLED RUBBER BAND

CONCAVE	PLANAR	CONVEX

6. How does each surface shape affect the size of the paper clip image as you look into the top of the test tube?

7. Examine a hand lens: How does its shape affect image size?

18

RAY PLAY ○ **Light ( )**

1. Cover the end of a flashlight with aluminum foil. Cut a straight, narrow slit (about $\frac{1}{3}$ across) that reaches to the rim but stops short of the center.

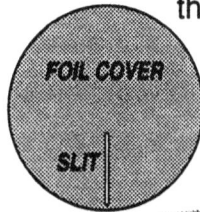

FOIL COVER

SLIT

2. Lay the flashlight, so that light shining through the slit casts a bright, narrow beam across white paper. (A dark work area helps you to see this most clearly.)

3. Interrupt this beam (near the slit) with each material below, experimenting with different angles. Report your findings in words and pictures.

    a. A small unframed mirror with straight sides.

    b. A sandwich of four microscope slides.

    c. A transparent pill vial (with and without water.)

19

THE COLOR SPECTRUM ○ **Light ( )**

1. Set a small jar on the bottom of a can. Rest both off center on a white plate.

2. Direct a source of light across the top of the jar at a sharp angle so the can's shadow falls across the plate.

3. Fill the jar with water until it brims above the lip. The rounded edge of water should refract a *spectrum* of color onto the shadowed plate.

4. List all the colors in this spectrum, starting farthest from the jar and working in.

LIGHT

ROUNDED WATER EDGE

SPECTRUM

5. Red has the longest wavelength of all the colors; violet the shortest. How does wavelength affect a color's tendency to refract?

6. Is white light a pure color?

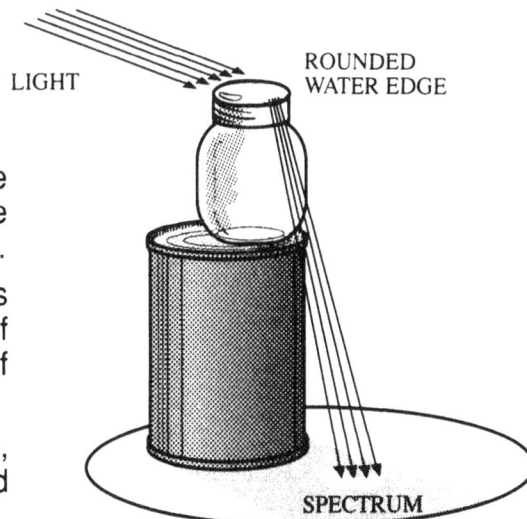

20

ADDITION AND SUBTRACTION ○   Light ( )

1. Most colored objects absorb and transmit a range of colors. Blue and yellow cellophane, for example, absorb and transmit these colors:

BLUE CELLOPHANE

YELLOW CELLOPHANE

a. Predict the color you get by passing white light through *both* blue and yellow cellophane. Explain.

b. Predict the color you get by *mixing* blue and yellow light on white paper. Explain.

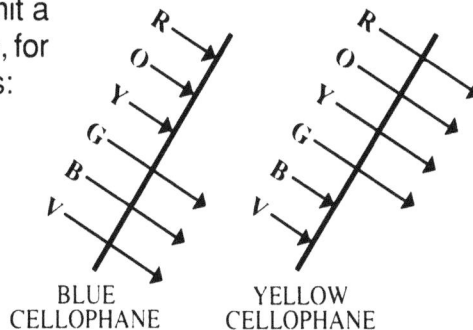

2. Rubber band some yellow cellophane around a mirror, letting it stick out at one end. Do a second mirror in blue. Experiment with sunlight to test your predictions.

SUNLIGHT
COLOR?
COLOR?

3. Yellow + blue yields two different colors. How can this be?

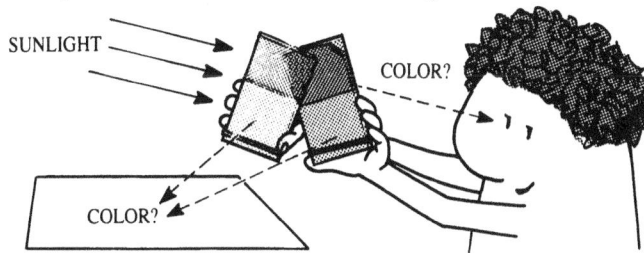

© 1991 by TOPS Learning Systems

21

FOCAL LENGTH (1) ○   Light ( )

1. Crease 2 sheets of notebook paper on the red margin lines. Wrap them around 2 books of equal size so each fold defines the top edge of each spine. Hold with rubber bands.

HALF OF LENS SHOWS
RUBBER RANDS
RED LINES, SPINES PARALLEL

2. Press the book spines together, then sink exactly half of a hand lens between them. Keep the spines parallel.

3. Darken the edge of an index card with the side of your pencil. Align this dark edge with each blue line you see through the lens, then draw along the card.

a. Your lens *converges* parallel lines to a *focal point*. What does this mean?
b. Measure the *focal length* — the distance from the focal point to the center of the lens.
c. Why do parallel "light rays" refract more towards the edges of the lens and less towards the center?

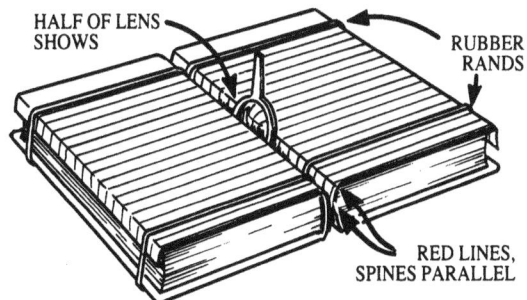

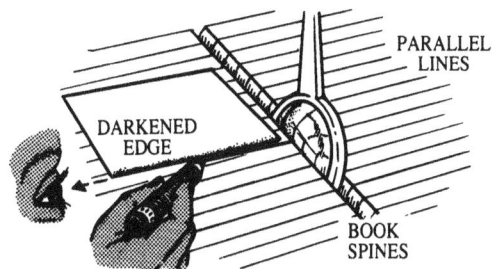

PARALLEL LINES
DARKENED EDGE
BOOK SPINES

© 1991 by TOPS Learning Systems

22

FOCAL LENGTH (2) ◯ Light (   )

1. Bend a paper clip at a right angle. Tape the narrower end to the handle of your lens like this:

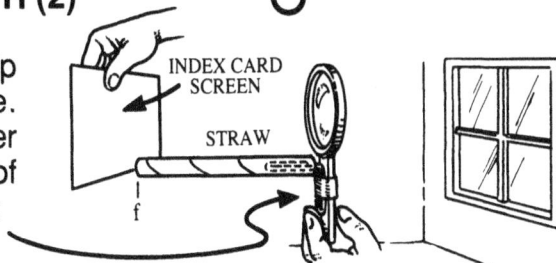

INDEX CARD SCREEN
STRAW
f

2. Push the wider end into a straw. Cut it to the focal length you determined before.

3. Hold your lens 1 focal length from an index card while facing it toward a distant scene outside a window. What do you see? (If the image isn't sharpest at the end of the straw, revise your accepted focal length.)

4. Make a jar of water slightly cloudy with a tiny pinch of powdered milk. Rubber band your straw (with lens) to the side of the jar, then reflect sunlight from a mirror or distant bright light down into the water. Block out excess light with a dark cloth or towel.

LIGHT
CLOUDY WATER
DARK CLOTH
RUBBER BAND

   a. Sketch the light cone you observe.
   b. Why does it focus beyond one focal length?

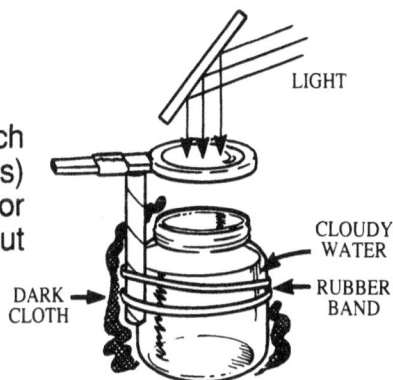

23

RAY RULES ◯ Light (   )

1. Set your lens between books covered with notebook paper, as before. Draw the *principal axis* — the normal that extends through its exact center.

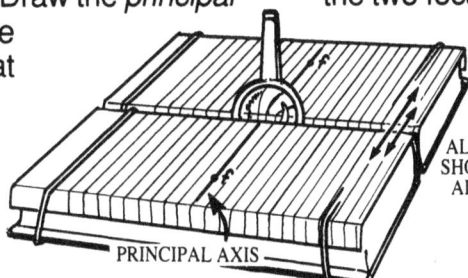

2. Align all lines on each paper. Measure and mark the two focal points on the principal axis.

ALL LINES SHOULD BE ALIGNED
PRINCIPAL AXIS

3. Draw 2 long straight pen lines on plastic wrap, then cut around each one.

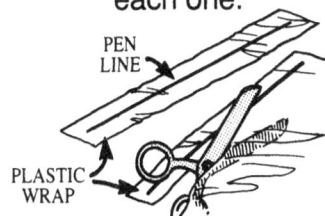

PEN LINE
PLASTIC WRAP

4. Line up both plastic "light rays" through the lens to complete each statement below. Illustrate each answer with a diagram that contains at least 2 rays. An incident light ray that:
   a. moves *parallel to the principal axis*, refracts…
   b. crosses the principal axis through its *focal point*, refracts along a path that…
   c. crosses the principal axis through the *center of the lens*…
   d. crosses the principal axis *inside the focal point*, refracts along a path that…
   e. crosses the principal axis *beyond the focal point*, refracts along a path that…

24

VIRTUAL IMAGES ○ Light ( )

1. Crease lined paper down the middle. Draw a *principal axis* along this fold. Outline a side view of your hand lens (about actual size) in the middle, and draw its *vertical axis*.

2. Measure and label the 2 focal points. Draw an observer's eye to the far left.

3. Tape a pin to the principal axis, tilting it as shown. Position its center about one third focal length from the lens.

4. Sketch how rays from the top and bottom of the pin should refract through the lens; then draw where the observer's eye should see them. Justify your drawing.

5. Notch out your lens drawing so a real lens can be inserted, between books. Bob your head up and down between the lens image and its predicted position. Comment on the accuracy of your drawing.

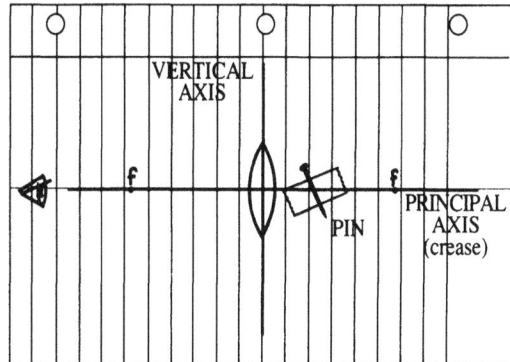

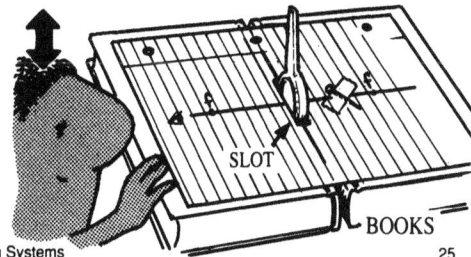

25

REAL IMAGES ○ Light ( )

1. Cut notebook paper in half lengthwise. Tape the sections end to end.

2. Draw your lens, and a long principal axis down the middle. Measure out 3 focal lengths to each side. Draw in the vertical axis.

3. Tape the points of 2 pins so they touch the principal axis where shown.

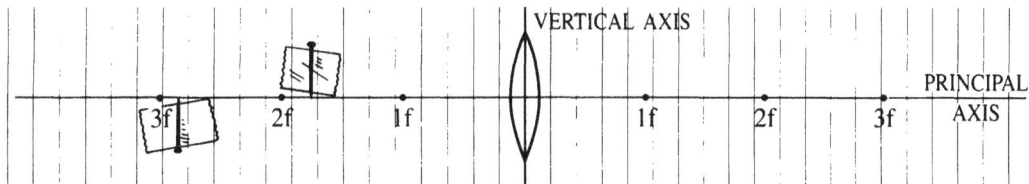

4. Locate the image of each pin head by drawing rays. Draw the rest of each pin to its point on the principal axis.
   a. Describe each pin image.
   b. Tape a pin to the front of your flashlight. Confirm that each image shows up at the predicted distance, by projecting its image through your hand lens onto an index card.

26

FLIP-FLOP   O   Light (   )

1. Stick a strip of masking tape to the side of a cereal box. Calibrate this tape in focal lengths (of your hand lens), starting at the bottom of the box.

2. Cut out the "object arrows" rectangle. Fold it along the dashed lines and tape it to the bottom of the box, so the arrows extend beside the calibrated tape.

CALIBRATED TAPE

OBJECT ARROWS

CEREAL BOX

3. Move the hand lens through these focal length positions while looking through it, *at arm's length*, with one eye. Fully describe each arrow image you see, and draw a ray diagram.

  a. from 0 f.l. to 1 f.l.
  b. at 1 f.l.
  c. from 1 f.l. to 2 f.l.
  d. at 2 f.l.
  e. beyond 2 f.l.

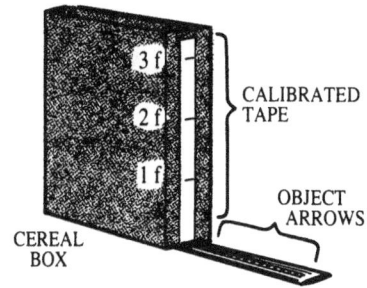

HOLD THE BOX AT *FULL* ARM'S LENGTH

4. Estimate the power of magnification of your lens: how many times larger can it make the arrows? Explain how you know.

5. Real images float right in space! Capture these arrows on waxed paper. Where should you hold it to see an actual-sized image?

© 1991 by TOPS Learning Systems   27

LENSES IN COMBINATION (1)   O   Light (   )

1. Rubber band waxed paper over your flashlight, and tape a paper clip on the front surface. Mount this on 2 cans with more rubber bands.

2. Bend a paper clip at a right angle. Tape the narrow end to the handle of your lens. Attach a straw cut to the focal length of your lens. Mount it on a battery like this:

|← 1 f.l. →|

3. Mount waxed paper on another battery. Flatten the top to make a screen, as shown.

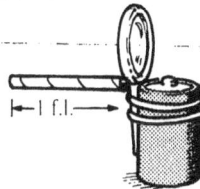

FLATTEN

4. Project an image of the paper clip through the lens onto the screen. Magnify the *back* of this screen image with another hand lens.

  a. Describe the first image on the screen; the second that you magnify.
  b. Can you remove the screen and still see this second image? Explain.

5. How should you arrange your equipment...
  a. ...to model a microscope (make the paper clip very large).
  b. ...to model a telescope (see the paper clip at a distance).
          (Save your lens with straw attached for the next activity.)

© 1991 by TOPS Learning Systems   28

---

cards 27-28

LENSES IN COMBINATIONS (2) ○   Light (   )

1. Tape the narrow ends of bent paper clips to 4 hand lenses like these. (Note: 2 lenses are taped together.)

SINGLE LENS
(1 paper clip)

SINGLE LENS
(2 paper clips)

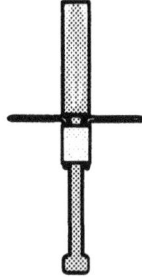

DOUBLE LENS
(1 paper clip)

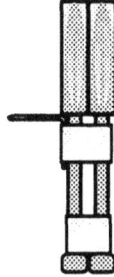

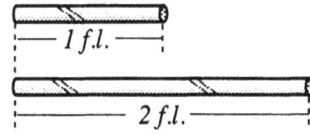

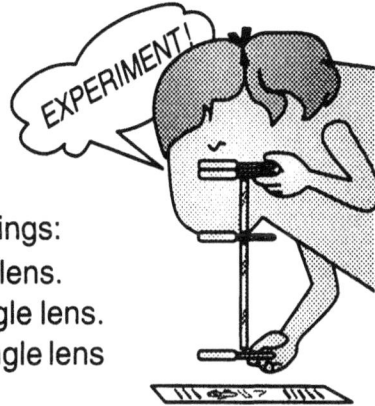

2. Cut straws to these focal lengths.

— 1 f.l. —

2 f.l.

EXPERIMENT!

3. Connect these combinations. Report your findings:

   a. **Inverter:** single lens + long straw + single lens.

   b. **Telescope:** double lens + short straw + single lens.

   c. **Microscope:** double lens + short straw + single lens
     + long straw + single lens.

© 2000 by TOPS Learning Systems

29

WHERE IS THE FOCUS? (1)   ○   Light (   )

1. Rubber band plastic wrap over a canning ring. Pull on the edges to make it wrinkle free, then deposit a line of different-sized water drops *inside* with a paper clip.

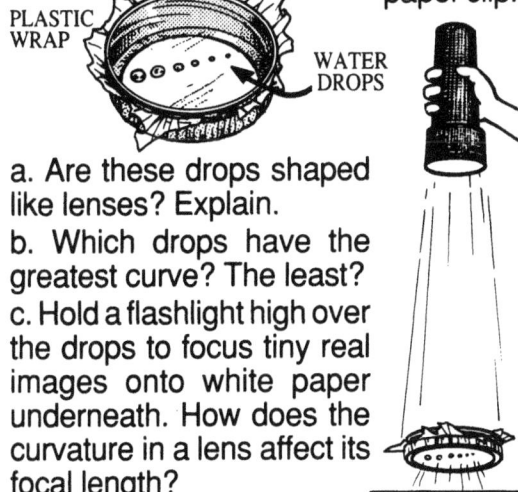

PLASTIC WRAP

WATER DROPS

a. Are these drops shaped like lenses? Explain.

b. Which drops have the greatest curve? The least?

c. Hold a flashlight high over the drops to focus tiny real images onto white paper underneath. How does the curvature in a lens affect its focal length?

2. To clearly see this point at arm's length, the eye must focus its diverging rays to a corresponding point on your retina.

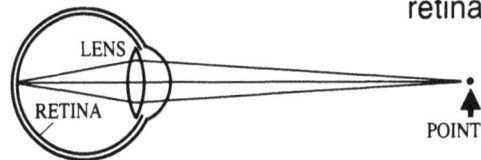

LENS

RETINA

POINT

a. Can you see this same point clearly only a hand-span away? Redraw the diagram, to explain how the eye accommodates (refocuses) to this shorter distance.

POINT

b. Can you see this point clearly only a thumb-width away? Redraw the eye diagram to illustrate.

© 1991 by TOPS Learning Systems

30

WHERE IS THE FOCUS? (2) O   Light (   )

1. Rubber band waxed paper to the front of your flashlight. Fix a pin to it with clear tape.

2. If you wear glasses, remove them. Hold a hand lens directly against your eye while looking at the lighted pin held at arm's length.
    a. How does it look?
    b. Draw a ray diagram that shows why this lens makes you near-sighted.

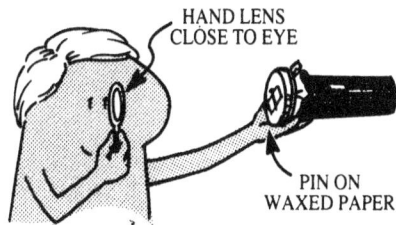

HAND LENS CLOSE TO EYE

PIN ON WAXED PAPER

3. Poke a *very* tiny hole with the tip of a pin in a small piece of foil. Press this hole between the hand lens and your eye. Now look at the lighted pin at arm's length, as before.
    a. How does the pin look?
    b. Redraw your ray diagram from step 3, to show why the pinhole refocuses the fuzzy lens image.

FOIL WITH TINY HOLE

4. Near-sighted people sometimes squint through nearly closed eyes to improve their distance vision. Try this while pressing the hand lens to your eye. Explain why this helps.   (Keep your flashlight covered for the next activity.)

31

THREE MAGNIFIED VIEWS O   Light (   )

1. Slowly bring the pin (taped to your lighted flashlight) in from arm's length closer to your eye.

a. Why does it look larger and larger?
b. Why does the image eventually blur?
c. Could you see the pin more clearly at close range through a pinhole? Make a prediction.
d. Test your prediction with different sized holes.

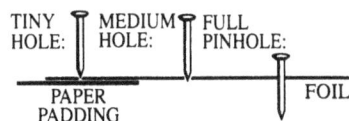

TINY HOLE: MEDIUM HOLE: FULL PINHOLE:

PAPER PADDING   FOIL

2. Rubber band plastic wrap over a canning ring, as before, so it is wrinkle free. Deposit drops of various sizes on the *outside* with a paper clip, then look through them at your lighted pin.

a. Which drops magnify the most?
b. Are the images in steps 1 and 2 real or virtual? Explain.

3. Remove the waxed paper and tape the pin directly to the flashlight head. Hold the hand lens *just beyond* 1 focal length, projecting the pin image onto an unlit wall.

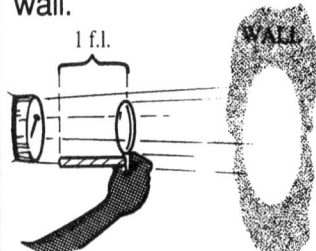

1 f.l.

WALL

a. How can you change the magnification?
b. Compare images in steps 2 and 3.

32

REVERSE IMAGES ○   Light (   )

1. Write your name or initials in large capital letters between two lines of notebook paper.

2. Hold a microscope slide *upright* on the top line. Look *down* through the slide at a steep angle, tracing the reflection on the other side.

3. Move the slide up one line at a time, drawing reflections of reflections, until you complete four lines. What pattern have you created?

MICROSCOPE SLIDE

4. Poke your pencil through a circle of index card so you can see the point only in a mirror.

INDEX CARD SHIELD

MIRROR

    a. Write your name so its mirror image appears normal.
    b. Why is this so hard to do?

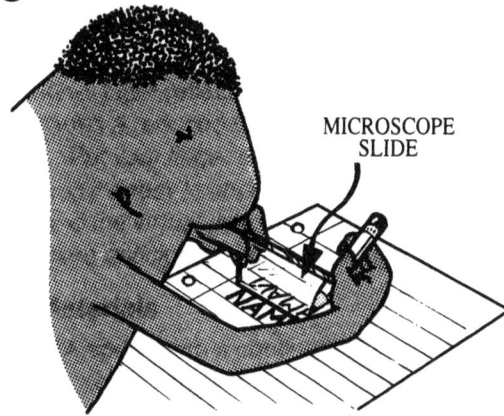

33

SYMMETRY ○   Light (   )

1. Use a mirror to find which letters are symmetrical — can be divided into similar halves. Summarize your findings in the table below, using each letter just once.

"A" has symmetry about its vertical axis, but not about its horizontal axis.

A B C D E F G H I J K L M
N O P Q R S T U V W X Y Z

no symmetry	
vertical axis of symmetry only	
horizontal axis of symmetry only	
vertical and horizontal axes of symmetry	

2. Draw a letter "X" that has *four* axes of symmetry. How did you do this?

3. Alter the letter "O" to make over a million axes of symmetry! Explain.

4. Spell a word that has horizontal line symmetry; vertical line symmetry.

34

CURVED MIRRORS (1) ◯ Light (   )

1. First draw a *semicircle* with a *principal axis*, and label the *center* of curvature C. This represents a concave mirror.

2. Next stand an object *arrow* on the axis beyond C, and extend a *parallel ray* from its tip to a *point* on the mirror.

3. Draw a *normal* through C to this point, and a *tangent* perpendicular to the normal.

4. Think of this tangent as a plane mirror. Draw the *reflected* ray using a protractor. Label the *focal length* where it intersects the principal axis.

5. Draw another ray from the tip of the object arrow through C. Because this ray reflects back on itself, you can locate the tip of the image arrow. Describe it.

6. Repeat steps 1-5 for a convex mirror of similar size. Describe the resulting image.

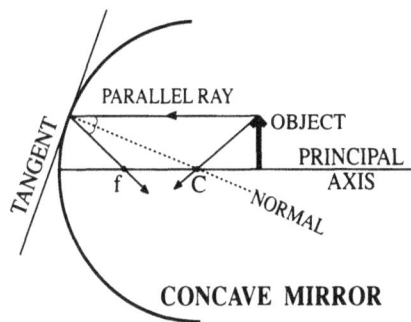

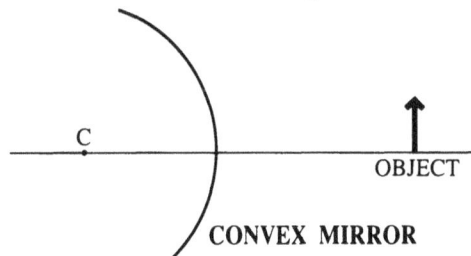

PARALLEL RAY
OBJECT
PRINCIPAL AXIS
TANGENT
f C
NORMAL

CONCAVE MIRROR

C
OBJECT

CONVEX MIRROR

35

CURVED MIRRORS (2) ◯ Light (   )

1. Wrap a spoon in a *single* layer of foil, shiny side up. Trim, leaving a small margin to anchor around the edge.

2. Vigorously polish the foil, front and back, with tissue, so you can easily see the reflection of your head and shoulders.

   a. Is your image on the concave side as your previous drawing predicts? Explain.
   b. What about the convex side?

NARROW MARGIN
SHINY FOIL

3. Cup the concave side of the spoon over your eye. This places your eye *inside* the spoon's focal length. Let in enough light from the side to view the image.

   a. Describe what you see.
   b. Draw a ray diagram to show how this image forms.

4. Hold a shining flashlight in front of a hand lens like this:

   a. Are the two "floating" images you see refractions or reflections? Why?
   b. Which is real and which is virtual? Use waxed paper to find out.
   c. How are these images formed?

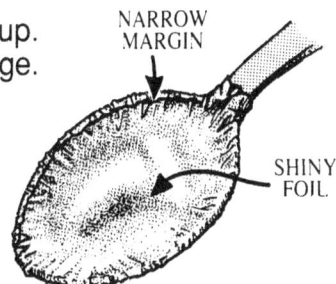

36

Supplementary Page

NORMAL

0°
10 10
20 20
30 30
40 40
50 50
60 60
70 70
80 80

MIRROR   BASELINE

Tape knot here:

OBJECT ARROWS:

fold back

↑↑

fold back

0 cm
1
2
3
4
5
6
7
8
9
10
11
12
13
14
15
16
17
18
19
20

Feedback

If you enjoyed teaching TOPS please tell us so. Your praise motivates us to work hard. If you found an error or can suggest ways to improve this module, we need to hear about that too. Your criticism will help us improve our next new edition. Would you like information about our other publications? Ask us to send you our latest catalog free of charge.

For whatever reason, we'd love to hear from you. We include this self-mailer for your convenience.

Ron and Peg Marson

author and illustrator

Your Message Here:

Module Title _____ Date _____

Name _____ School _____

Address _____

City _____ State _____ Zip _____

————————————————————————— FIRST FOLD —————————————————————————

————————————————————————— SECOND FOLD —————————————————————————

RETURN ADDRESS

———————————————————————

———————————————————————

———————————————————————

PLACE
STAMP
HERE

TOPS Learning Systems
342 S Plumas St
Willows, CA 95988

TAPE HERE

www.ingramcontent.com/pod-product-compliance
Lightning Source LLC
Chambersburg PA
CBHW051118200326
41518CB00016B/2548